Underwater Grasses

in Chesapeake Bay & Mid-Atlantic Coastal Waters

Guide to Identifying Submerged Aquatic Vegetation

Peter W. Bergstrom, Robert F. Murphy, Michael D. Naylor, Ryan C. Davis, and Justin T. Reel

This guide is the result of a cooperative effort among the Maryland Sea Grant College, the National Oceanic and Atmospheric Administration Chesapeake Bay Office, the Alliance for the Chesapeake Bay, and the Maryland Department of Natural Resources. It is also based on work completed by RK&K Engineers and the Alliance for the Chesapeake Bay under a contract with the Department of Defense. The editors are grateful for the assistance of all these partners, and for the generous contributions each of them have made to this publication.

Published by Maryland Sea Grant College under grant NA 05OAR4171042 from the National Oceanic and Atmospheric Administration.

Maryland Sea Grant Publication Number UM-SG-PI-2006-01
Book and cover design by Sandy Rodgers

This book may be ordered from: Maryland Sea Grant College, 4321 Hartwick Road, Suite 300, College Park, MD 20740
or from the online bookstore at www.mdsg.umd.edu

Library of Congress Cataloging-in-Publication Data

Underwater grasses in Chesapeake Bay & Mid-Atlantic coastal waters : guide to identifying submerged aquatic vegetation / Peter W. Bergstrom ... [et al.].
p. cm.
Includes bibliographical references.
ISBN-13: 978-0-943676-64-7 (alk. paper)
ISBN-10: 0-943676-64-9 (alk. paper)
1. Aquatic plants—Chesapeake Bay (Md. and Va.)—Identification. 2. Aquatic plants—Middle Atlantic States—Identification. 3. Aquatic plants—Atlantic Coast (U.S.)—Identification. 4. Seagrasses—Chesapeake Bay (Md. and Va.)—Identification. 5. Seagrasses—Middle Atlantic States—Identification. 6. Seagrasses—Atlantic Coast (U.S.)—Identification. I. Bergstrom, Peter Wasson, 1952-
QK122.8.U53 2006
581.7'60916347—dc22 2006009596

Acknowledgments

This guide would not have been possible without the hard work and support of a number of key individuals and organizations. *Underwater Grasses* builds on a prototype publication originally produced under a cooperative agreement between the U.S. Department of Defense and the Alliance for the Chesapeake Bay in conjunction with Rummel, Klepper & Kahl, LLP (RKK). Previous guides have also proven instrumental, including *Field Guide to the Submerged Aquatic Vegetation of Chesapeake Bay*, produced by the U.S. Fish and Wildlife Service. We are grateful for the use of photographs by Linda M. Hurley and others from that guide.

Thanks also to David Jasinski of the University of Maryland Center for Environmental Science for creating the Chesapeake Bay maps of salinity ranges in wet and dry years (p. 5) and to Paula Jasinski and Tara Shleser at the National Oceanic and Atmospheric Administration (NOAA) Chesapeake Bay Office for creating the maps of potential salinity ranges for the species of underwater grasses included in this guide (pp. 67-70).

The authors would like to thank Jack Greer, Sandy Rodgers, and Erica Goldman of Maryland Sea Grant for overseeing editing, design, and production of the guide, and Vicki Meade of Meade Communications for editorial assistance. They also acknowledge the financial support of NOAA, Sea Grant, and the Alliance for the Chesapeake Bay, and a generous grant from Constellation Energy.

The authors are grateful for the careful review of the manuscript by Steven Ailstock of Anne Arundel Community College, Tom Parham of the Maryland Department of Natural Resources, and Kathryn Reshetiloff of the U.S. Fish and Wildlife Service Chesapeake Bay Field Office.

TABLE OF CONTENTS

CONTENTS BY COMMON NAME

Underwater grasses, key components of many coastal ecosystems, provide habitat, food, and shelter. These plants — though once considered a nuisance by some — help keep coastal waters like Chesapeake Bay healthy by absorbing nutrients and filtering sediment. More accurately called "submerged aquatic vegetation" or SAV, these grasses enrich shallow aquatic environments around the world, providing sanctuaries for crabs and finfish, and sustenance for waterfowl.

SAV can grow in fresh, brackish, or sea water, and more than 500 species of SAV inhabit the world's rivers, lakes, estuaries, and oceans. Chesapeake Bay and its tidal tributaries are home to about 20 common species of SAV. A few of these species are not native but have become widely established throughout the Bay.

Types of Aquatic Vegetation

Submerged Aquatic Vegetation (SAV):

These plants usually grow entirely underwater and rely on buoyancy to support their stems and leaves. A few species have flowers or tufts that grow slightly above the water's surface. Like many land plants, SAV species are vascular, with leaf-stem-root systems. Unlike many land plants, their submerged parts lack a waxy cuticle or stomata and their flowers are often tiny and inconspicuous. Many species are annuals; each year before dying, they produce flowers and seeds from which new plants later grow. Others are perennials that flower and produce seeds, but survive for multiple years. Most SAV species, whether or not they flower, spread largely through asexual (vegetative) reproduction.

Floating Aquatic Vegetation (FAV):

These plants have leaves floating on the water's surface and may or may not be attached to the substrate by underwater roots or stems. Like SAV, they rely on water to support them, but they have a waxy cuticle and stomata on the upper surface of their floating leaves. FAV are mostly found in ponds and will not be described in this guide, with the exception of water chestnut, a species that has also become a troublesome invasive in tidal waters.

Emergent Aquatic Vegetation (EAV): These plants (not included in this guide) are commonly found in tidal marshes along the shores of the Bay and its tributaries. Unlike SAV, the upper stems and leaves of these plants have a waxy cuticle and protrude in the air without water supporting them; only the lower stems and roots are below the water surface. Common examples include cord grass (*Spartina* spp.) and sweet flag (*Acorus calamus*).

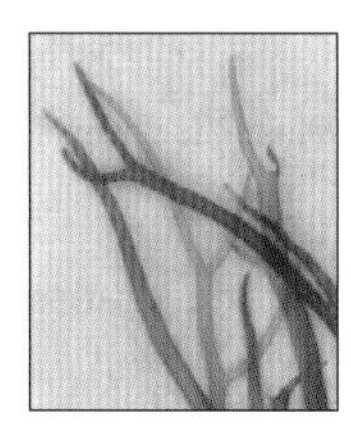

Algae: Algae are non-vascular plants, lacking the characteristic root-stem-leaf systems and specialized plant tissues that transport water and nutrients. Some algal species can be confused with SAV because they look similar and overlap in distribution. Some algae are phytoplankton — single-celled, free-floating plants. Others are multi-cellular units that appear as slimy green mats or clumps on rocks or pilings. Several species can be abundant in the Bay and are described briefly in this guide.

Blue-Greens (Cyanobacteria): Although more commonly known as blue-green algae, cyanobacteria are not plants at all. They are photosynthetic bacteria that occur in many aquatic and terrestrial habitats, including the polar ice caps. In Chesapeake Bay, cyanobacteria often bloom in summer months. Some species can release toxins that are harmful to human health. Cyanobacteria may appear similar to true algae, forming mats on the surface or long filamentous strands, but are not likely to be confused with SAV. A few species are discussed in this guide.

Using This Guide

This guide will help you identify common SAV species found in Chesapeake Bay and coastal waters of the Mid-Atlantic (pages with light green borders). The guide also includes information on some species of floating aquatic plants, algae, and cyanobacteria.

Designed for both the amateur and the expert, the guide includes a graphics-based key (p. 8) to help you determine which SAV species you find. A glossary in the back (p. 71) defines terms you may find unfamiliar. SAV species are arranged roughly from low to high salinity, though the salinity ranges of some plants may overlap. As shown in the sample page at right, each entry includes information on recognition, distribution, and reproduction. The lower border lists the salinity range in parts per thousand (ppt) that corresponds to the best growth condition for that species. Photographs will orient you to the appearance of the whole plant, with detailed pictures of structures, such as leaves or seeds, that will help with identification. A section on confusing species (pp. 48-52) will help to differentiate between species that can be easily mistaken for each other.

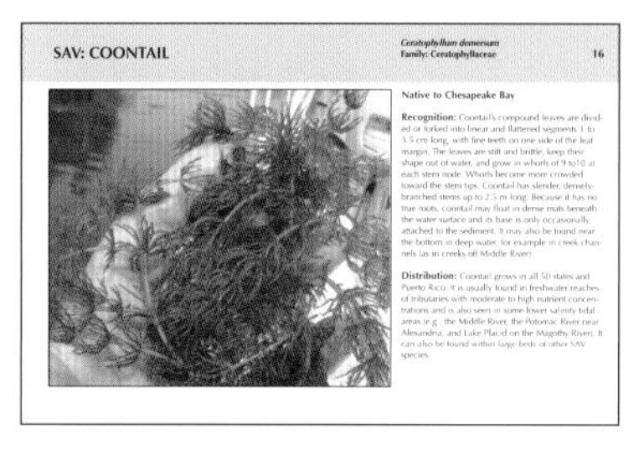

SAV: COONTAIL | *Ceratophyllum demersum* Family: Ceratophyllaceae | 16

Native to Chesapeake Bay

Recognition: Coontail's compound leaves are divided or forked into linear and flattened segments 1 to 3.5 cm long, with fine teeth on one side of the leaf margin. The leaves are stiff and brittle, keep their shape out of water, and grow in whorls of 9 to10 at each stem node. Whorls become more crowded toward the stem tips. Coontail has slender, densely-branched stems up to 2.5 m long. Because it has no true roots, coontail may float in dense mats beneath the water surface and its base is only occasionally attached to the sediment. It may also be found near the bottom in deep water, for example in creek channels (as in creeks off Middle River).

Distribution: Coontail grows in all 50 states and Puerto Rico. It is usually found in freshwater reaches of tributaries with moderate to high nutrient concentrations and is also seen in some lower salinity tidal areas (e.g., the Middle River, the Potomac River near Alexandria, and Lake Placid on the Magothy River). It can also be found within large beds of other SAV species.

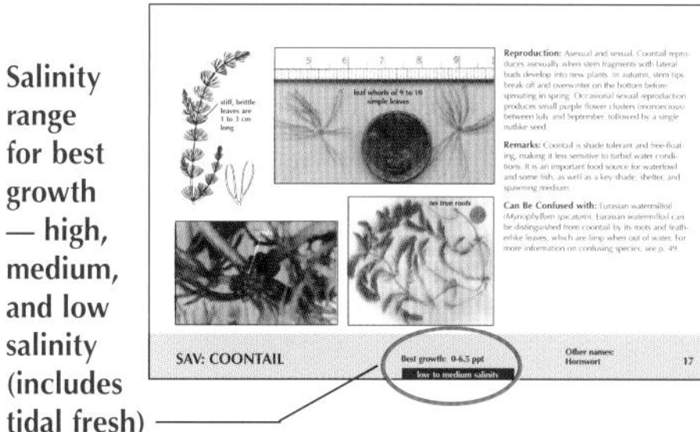

Reproduction: Asexual and sexual. Coontail reproduces asexually when stem fragments with lateral buds develop into new plants. In autumn, stem tips break off and overwinter on the bottom before sprouting in spring. Occasional sexual reproduction produces small purple flower clusters (monoecious) between July and September, followed by a single nutlike seed.

Remarks: Coontail is shade tolerant and free-floating, making it less sensitive to turbid water conditions. It is an important food source for waterfowl and some fish, as well as a key shade, shelter, and spawning medium.

Can Be Confused with: Eurasian watermilfoil (*Myriophyllum spicatum*). Eurasian watermilfoil can be distinguished from coontail by its roots and featherlike leaves, which are limp when out of water. For more information on confusing species, see p. 49.

SAV: COONTAIL | Best growth: 0-6.5 ppt low to medium salinity | Other names: Hornwort | 17

All underwater plants have adapted to a certain amount of salt — or lack of it — in their environment and salinity often determines where a particular species will grow. This guide organizes species according to the salinity ranges in which they are generally found — from the tidal fresh waters of Chesapeake Bay to full-strength sea water along the Mid-Atlantic coast, as far south as North Carolina.

In an estuary such as Chesapeake Bay, the amount of precipitation in the watershed will largely influence a given year's salinity pattern. In a wet year, when fresh water rushes down rivers into the Bay, salinity levels drop, especially in the northern Bay and up the tributaries. In a dry year, when less fresh water enters the system, salinity levels rise, especially in the mid to lower Bay. The maps below show salinity patterns for a representative dry year and a representative wet year in the Chesapeake.

A detailed table (pp. 6-7) lists the salinity range for each species of SAV featured in this guide, whether found in Chesapeake Bay or other Mid-Atlantic waters. The blue bar reflects the total salinity range over which the species can grow. The inset green bar reflects the salinity range for best growth.

To give you an idea of where SAV species have been found in recent years, you can see maps archived on the Virginia Institute of Marine Science (VIMS) web site at www.vims.edu/bio/sav/groundsurveymaps.html. Remember that surveys provide an overview, and distribution can change from year to year.

Salinity and Growth Range

A set of maps from the Chesapeake Bay Program (pp. 67-70) shows potential best growth range for a particular species based on salinity regime (sample map at right). These maps will help you identify which species is likely to be growing where. The maps were created using data from March to May mean surface salinity from 1985-2003.

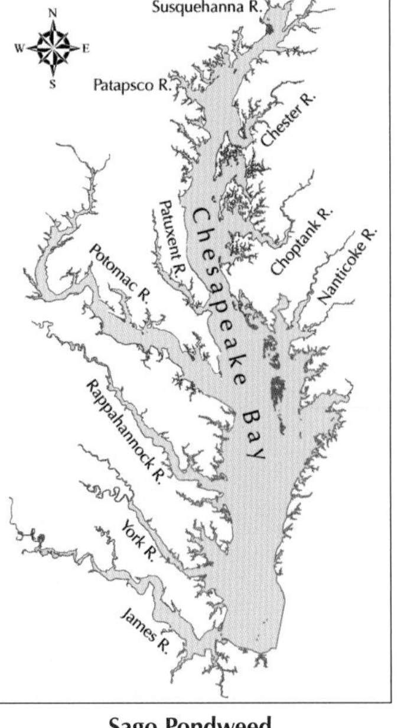

Sago Pondweed
(*Stuckenia pectinata*)
0 to 12 ppt
low to medium salinity

Salinity Ranges in Chesapeake Bay

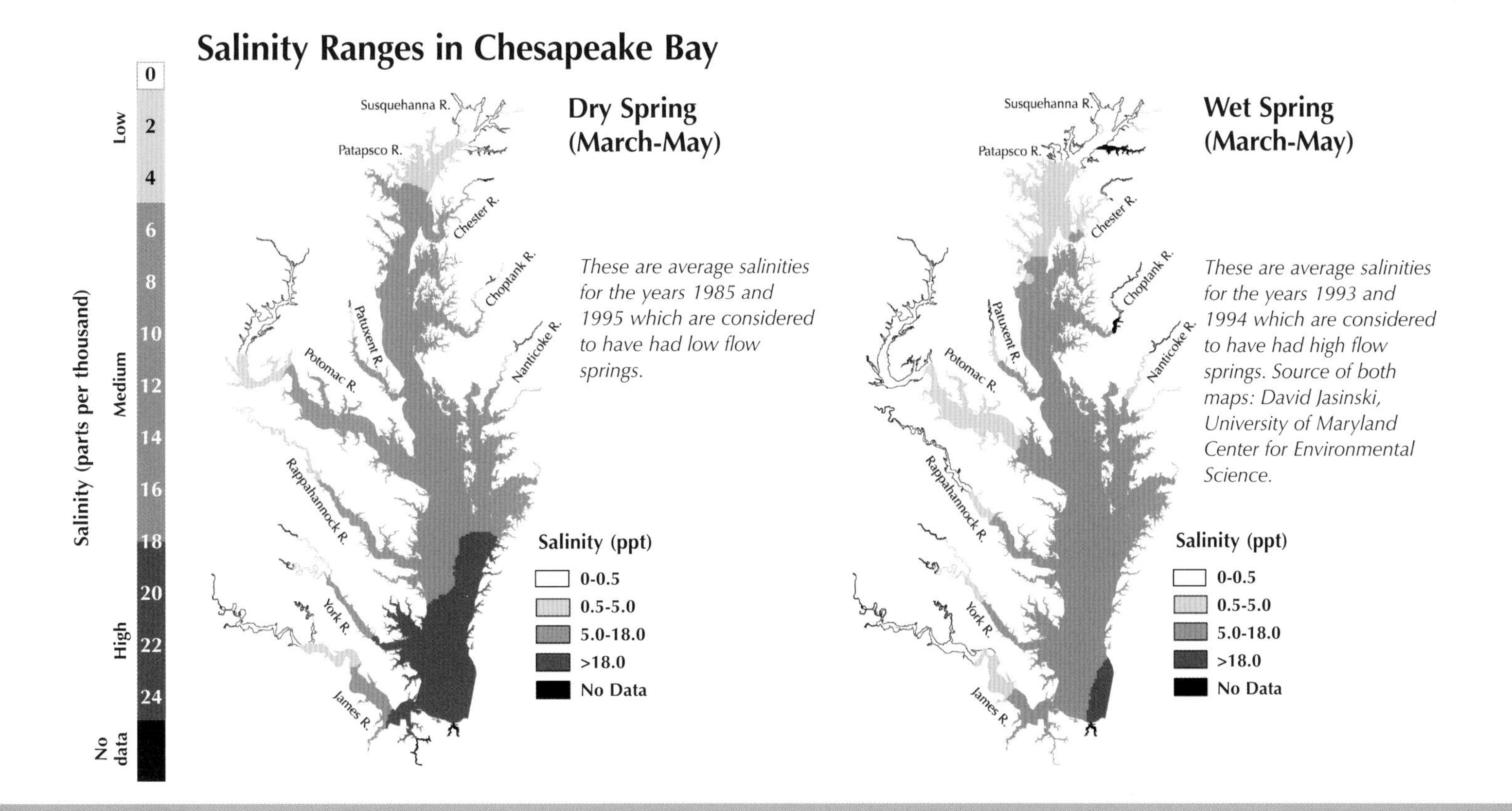

These are average salinities for the years 1985 and 1995 which are considered to have had low flow springs.

These are average salinities for the years 1993 and 1994 which are considered to have had high flow springs. Source of both maps: David Jasinski, University of Maryland Center for Environmental Science.

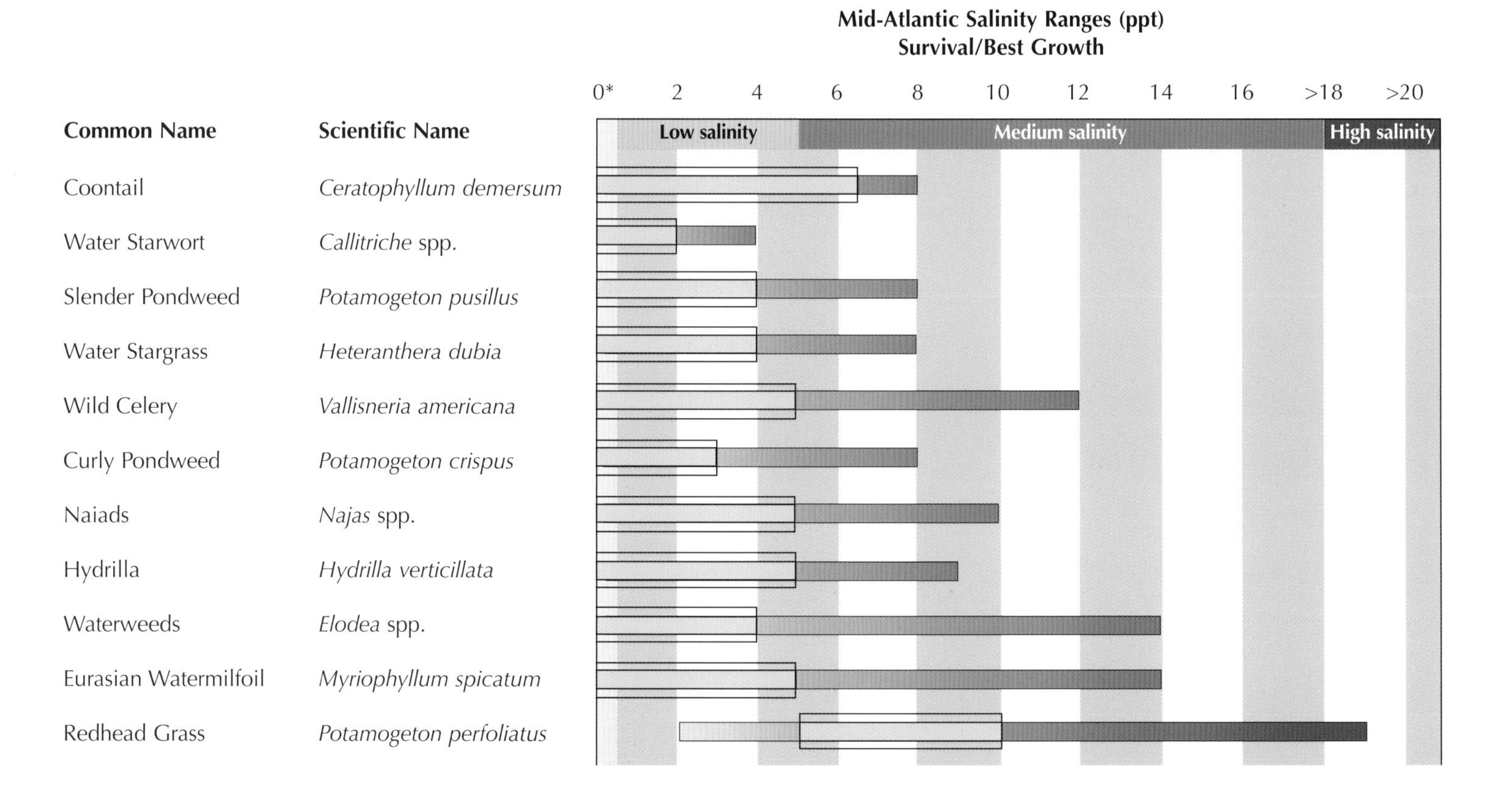
Mid-Atlantic Salinity Ranges (ppt)
Survival/Best Growth
0*
2
4
6
8
10
12
14
16
>18
>20
Common Name
Scientific Name
Low salinity
Medium salinity
High salinity
Coontail
Ceratophyllum demersum
Water Starwort
Callitriche spp.
Slender Pondweed
Potamogeton pusillus
Water Stargrass
Heteranthera dubia
Wild Celery
Vallisneria americana
Curly Pondweed
Potamogeton crispus
Naiads
Najas spp.
Hydrilla
Hydrilla verticillata
Waterweeds
Elodea spp.
Eurasian Watermilfoil
Myriophyllum spicatum
Redhead Grass
Potamogeton perfoliatus

Common name	Scientific name
Horned Pondweed	*Zannichellia palustris*
Sago Pondweed	*Stuckenia pectinata*
Widgeon Grass	*Ruppia maritima*
Eelgrass	*Zostera marina*
Shoal Grass	*Halodule wrightii*

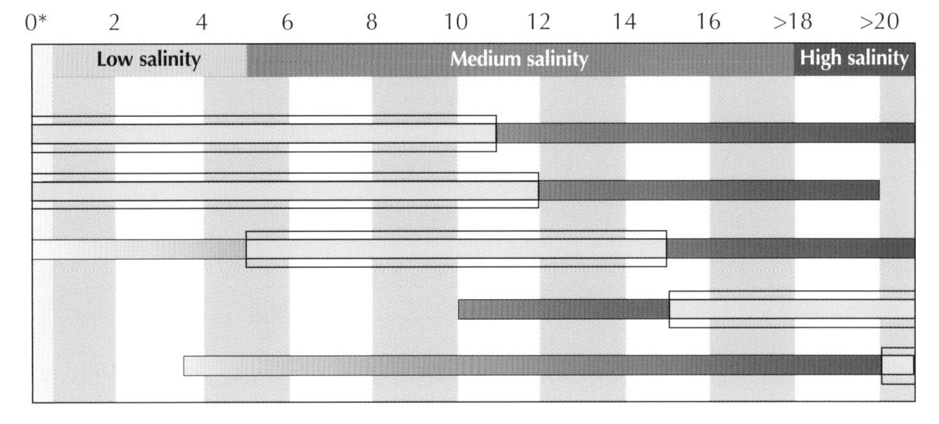

Species Salinity Ranges

*Tidal fresh (0-0.5 ppt)
Low salinity (0.5-5 ppt)
Medium salinity (5-18 ppt)
High salinity (>18 ppt)

Full growth range

Best growth range

Remarks: While the salinity ranges included in this table derive from a variety of sources, there are some known exceptions that are important to mention. Coontail (*Ceratophyllum demersum*), for example, grows in fresh water in aquariums and ponds but has not been found in fresh water in the Chesapeake. At certain salinities (5-14 ppt.), common waterweed (*Elodea canadensis*) may be confused with a very similar species, Nuttall's waterweed (*Elodea nuttallii*) — which is known to grow optimally in that range. Widgeon grass (*Ruppia maritima*) grows in nontidal fresh waters elsewhere at 0 ppt, though it is rarely found below 5 ppt in Chesapeake Bay. Many species can also withstand brief exposure to high salinities. Sources: Bergstrom, personal communication (2005); Stevenson and Confer (1978); Brown and Brown (1984); and Fonseca et al. (1998).

This key will allow you to identify most of the submerged aquatic vegetation species found in Chesapeake Bay and Mid-Atlantic coastal waters. Called a multichotomous key, it is set up to direct you through a series of choices that will lead you to a species or group of species. Each set of line drawings or text boxes allows you to choose between differentiating features. At the end of the decision process, a picture will direct you to the correct species or group of species. A hand lens, or magnifying glass, as well as a metric ruler with millimeter markings (see inside back cover) will aid in the successful identification of your unknown SAV. To begin, compare your unknown SAV with each of the four common leaf arrangements. Choose the picture which best represents your species and follow the arrows until you have properly identified your unknown species. Refer to the specified page number for a complete species profile. An online Bay Grass Key from the Maryland Department of Natural Resources provides more information: www.dnr.state.md.us/bay/sav/key.

Leaf Arrangement

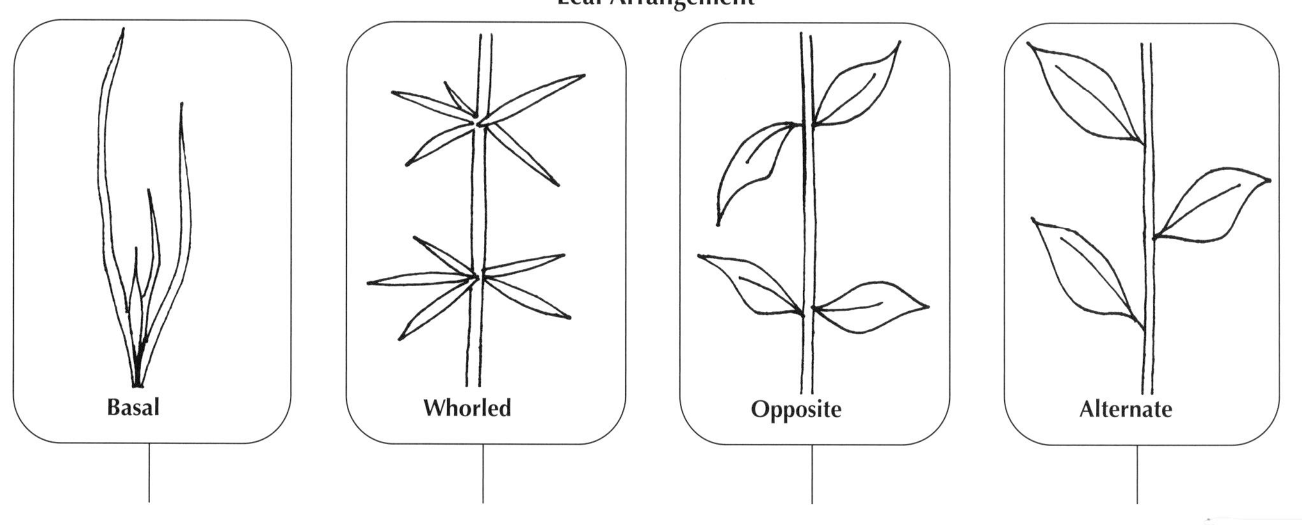

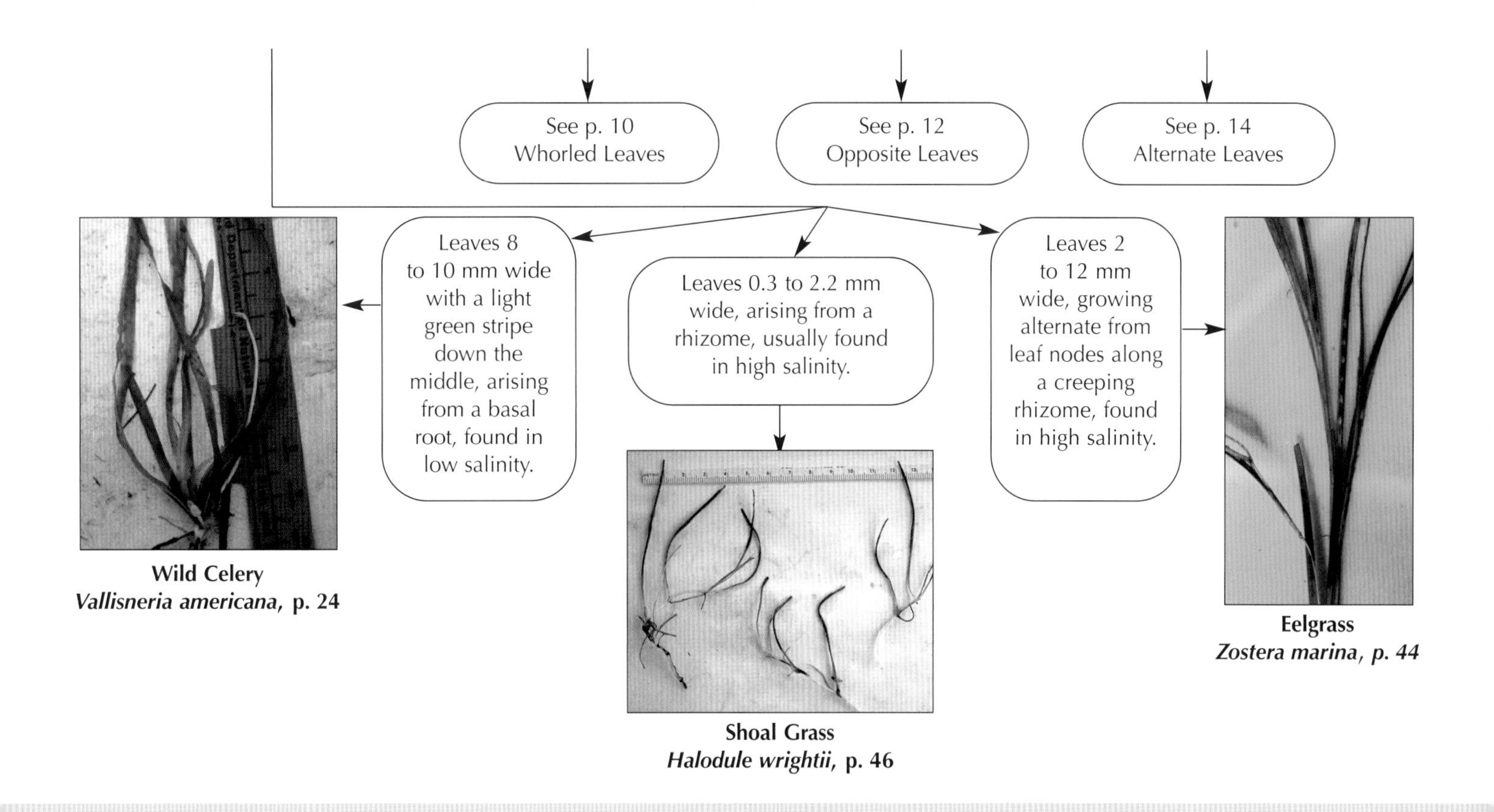

Wild Celery
***Vallisneria americana*, p. 24**

Shoal Grass
***Halodule wrightii*, p. 46**

Eelgrass
***Zostera marina*, p. 44**

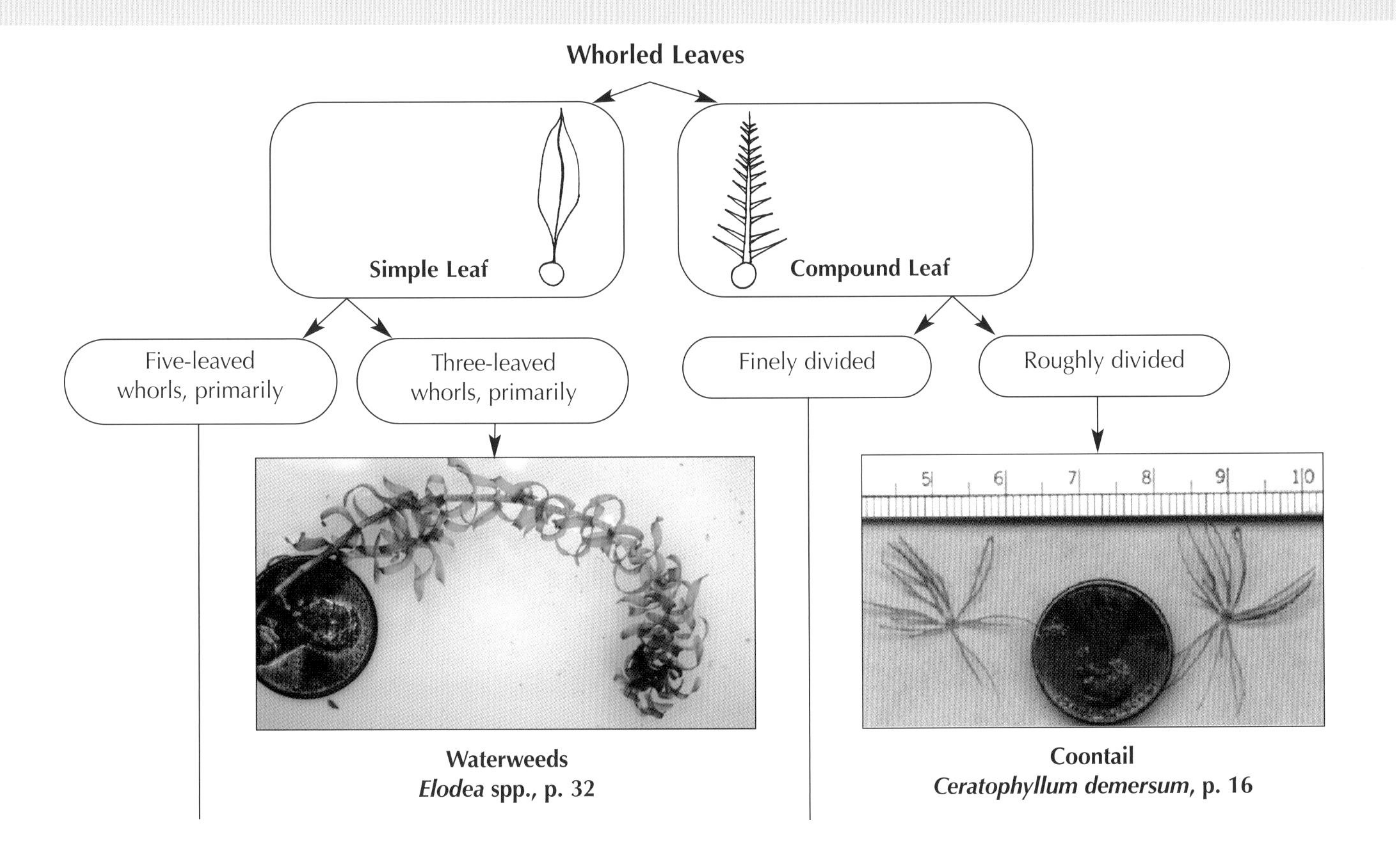

Waterweeds
Elodea spp., p. 32

Coontail
Ceratophyllum demersum, p. 16

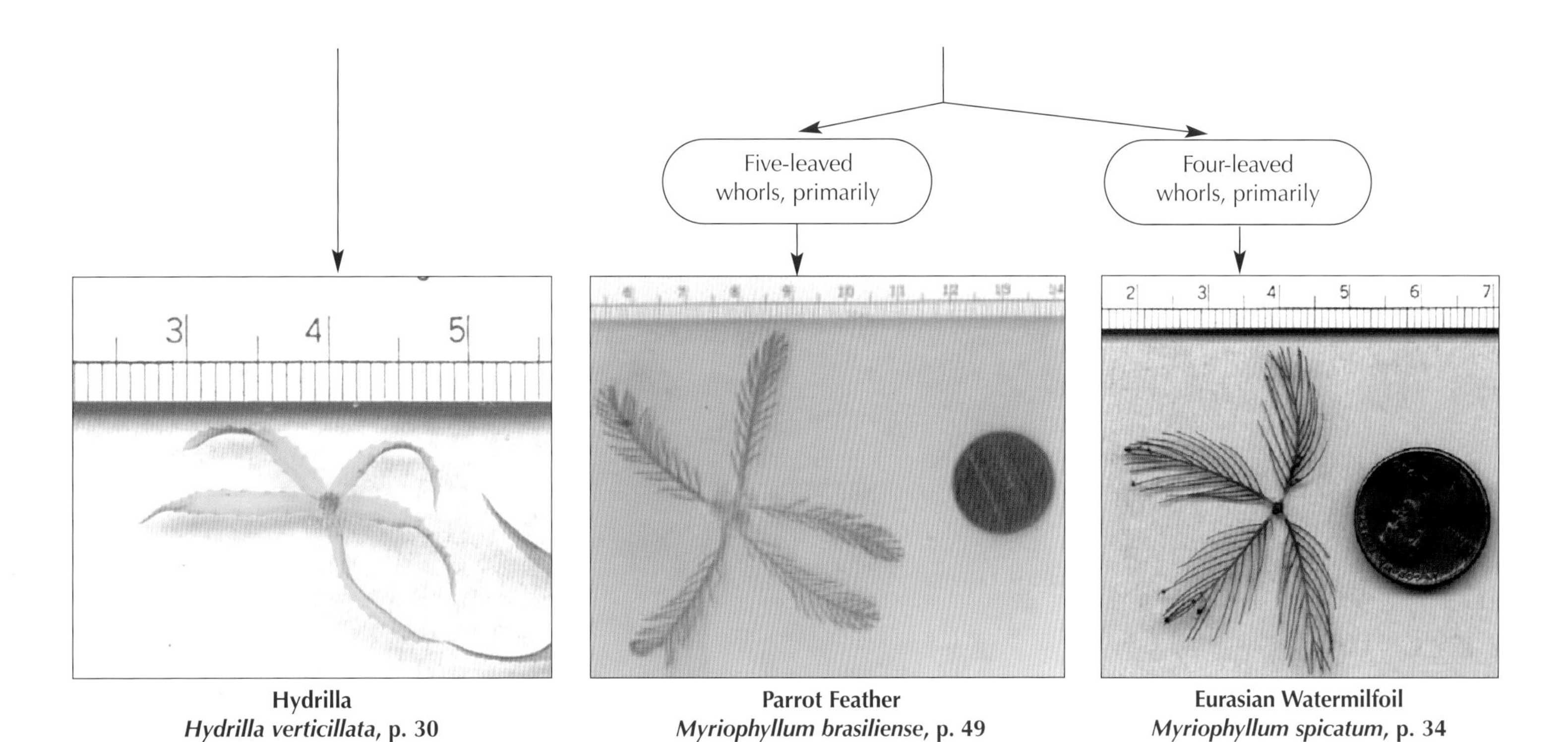

Hydrilla
Hydrilla verticillata, p. 30

Parrot Feather
Myriophyllum brasiliense, p. 49

Eurasian Watermilfoil
Myriophyllum spicatum, p. 34

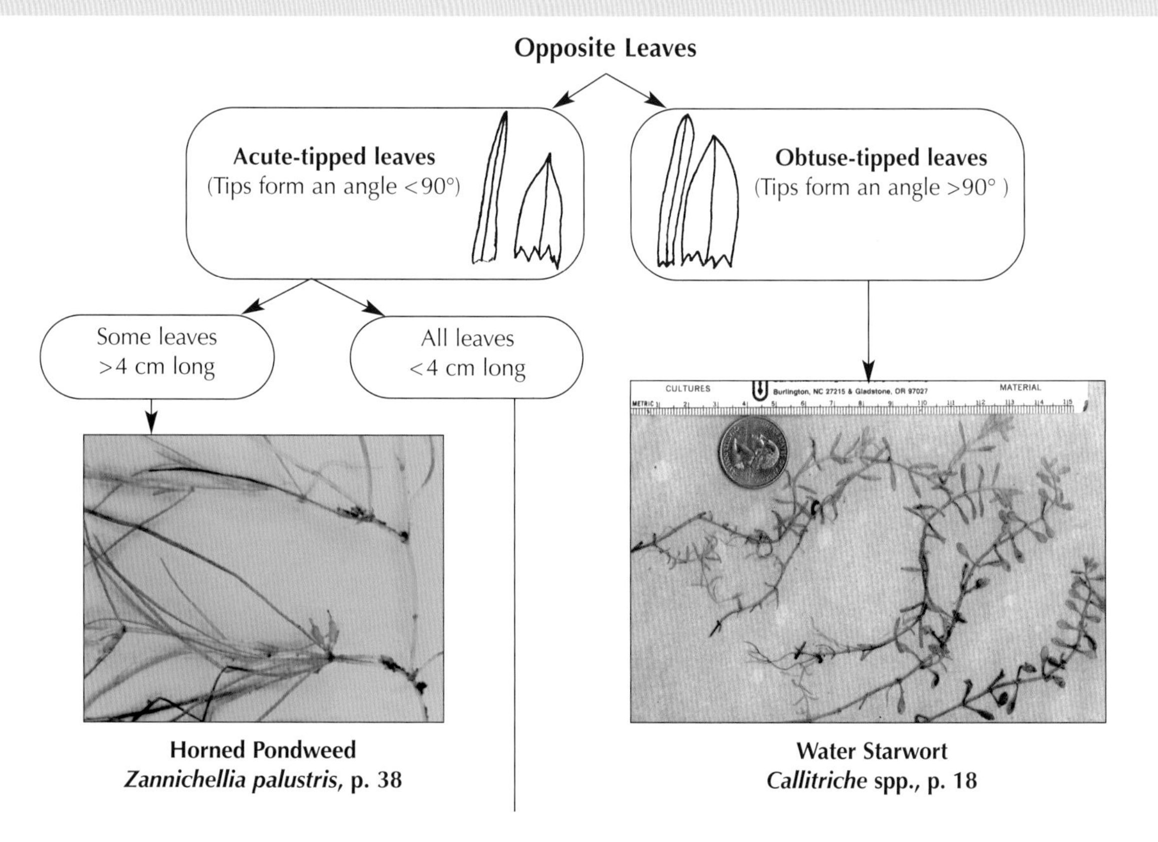

Horned Pondweed
***Zannichellia palustris*, p. 38**

Water Starwort
***Callitriche* spp., p. 18**

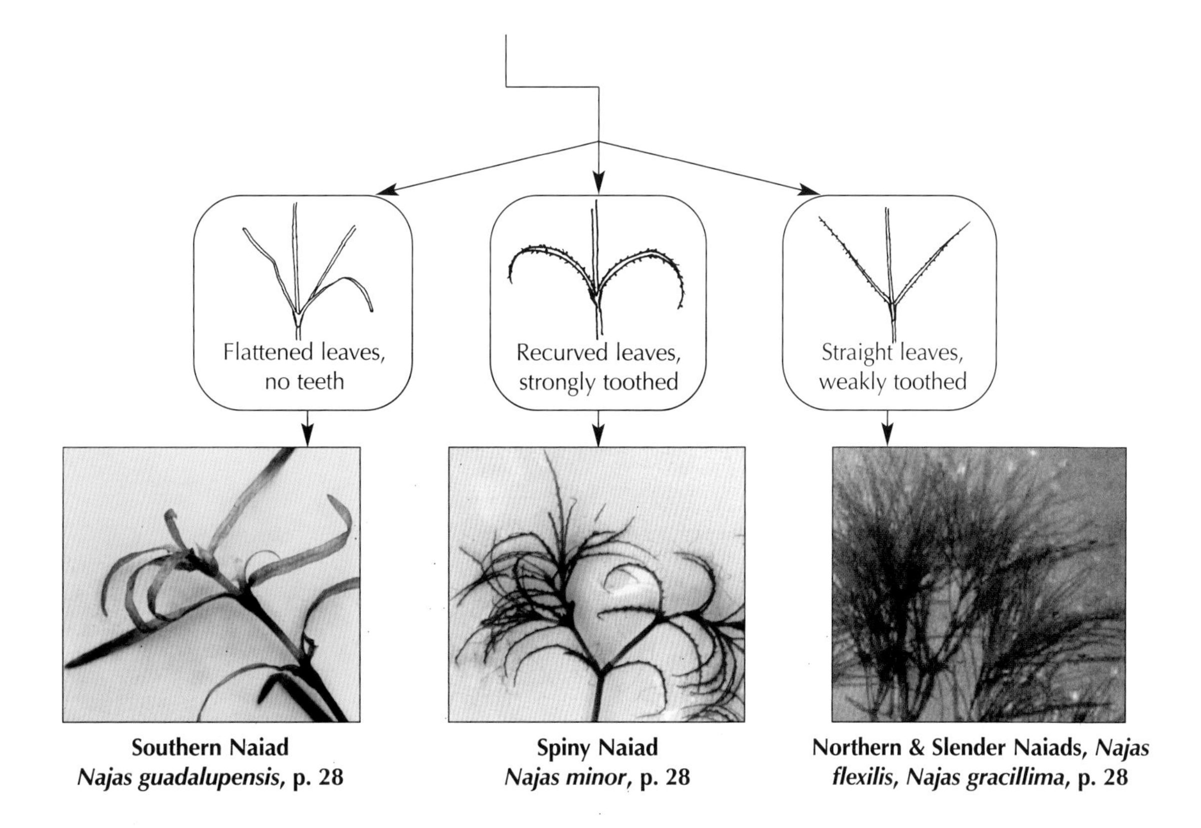

Southern Naiad
Najas guadalupensis, **p. 28**

Spiny Naiad
Najas minor, **p. 28**

Northern & Slender Naiads, ***Najas flexilis, Najas gracillima*****, p. 28**

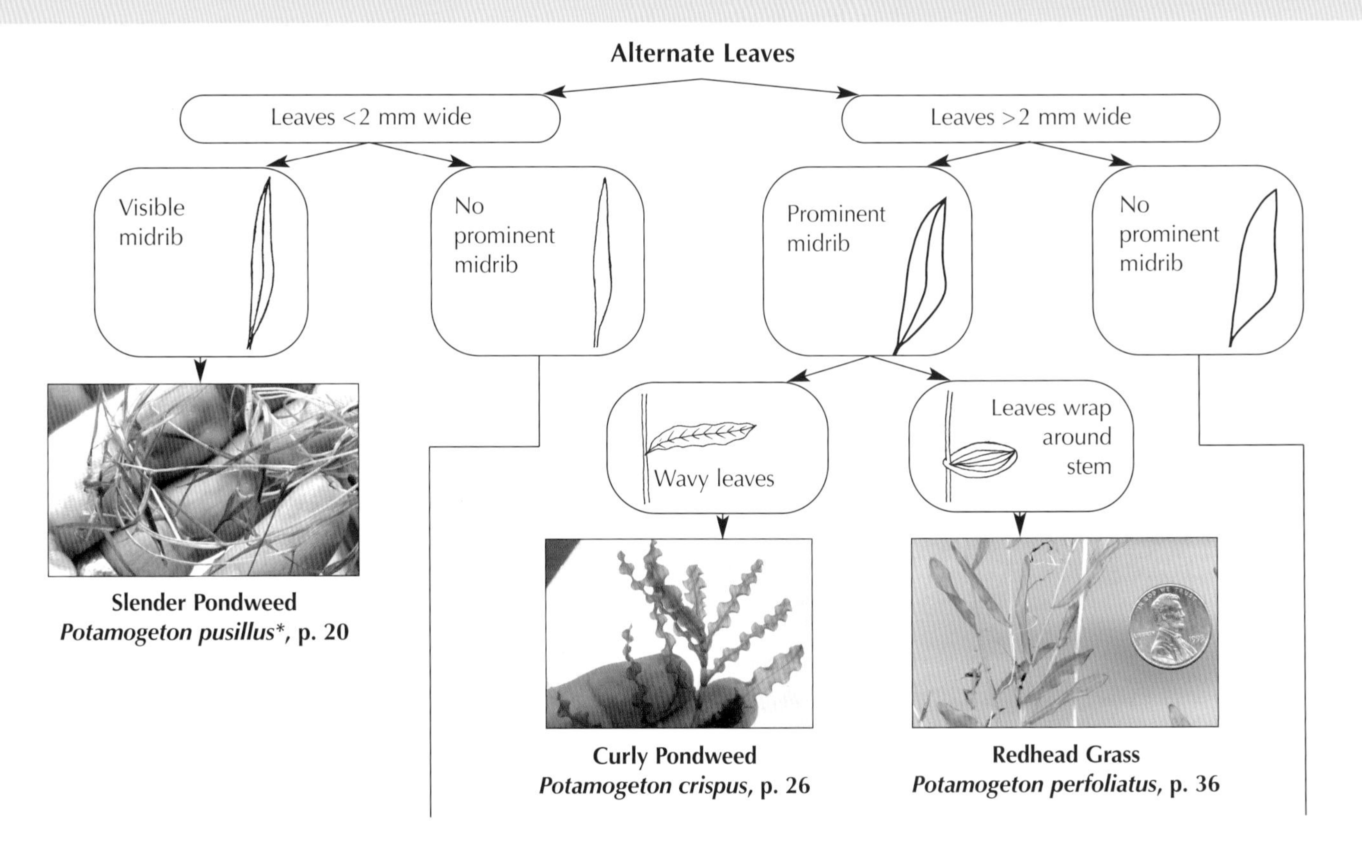
Alternate Leaves
Leaves <2 mm wide
Leaves >2 mm wide
Visible midrib
No prominent midrib
Prominent midrib
No prominent midrib
Wavy leaves
Leaves wrap around stem
Slender Pondweed
Potamogeton pusillus*, p. 20
Curly Pondweed
Potamogeton crispus, p. 26
Redhead Grass
Potamogeton perfoliatus, p. 36

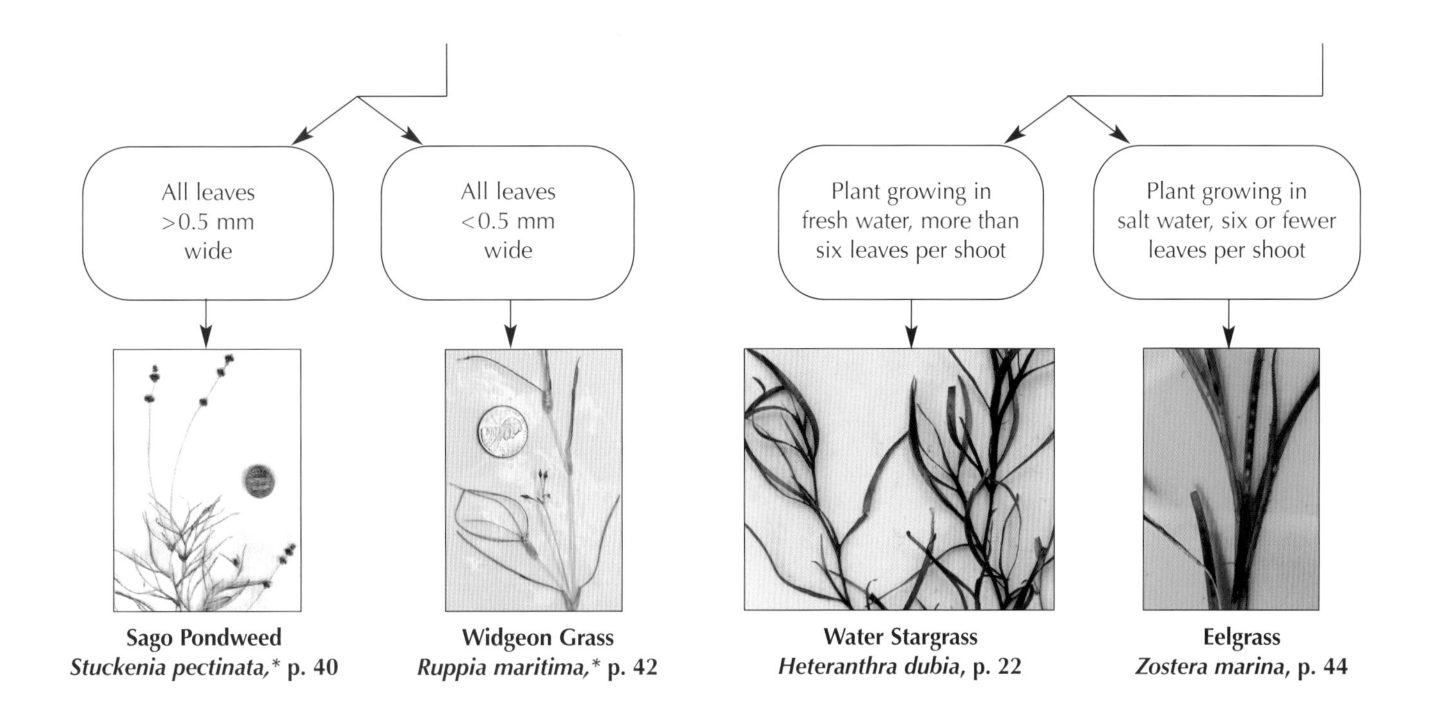

Sago Pondweed
***Stuckenia pectinata,** p. 40**

Widgeon Grass
***Ruppia maritima,** p. 42**

Water Stargrass
***Heteranthra dubia*, p. 22**

Eelgrass
***Zostera marina*, p. 44**

* All three alternate, narrow-leaved plants are very similar in appearance unless flowering or fruiting. Examine each species description carefully to determine the correct species.

SAV: COONTAIL

Ceratophyllum demersum
Family: Ceratophyllaceae

Native to Chesapeake Bay

Recognition: Coontail's compound leaves are divided or forked into linear and flattened segments 1 to 3.5 cm long, with fine teeth on one side of the leaf margin. The leaves are stiff and brittle, keep their shape out of water, and grow in whorls of 9 to 10 at each stem node. Whorls become more crowded toward the stem tips. Coontail has slender, densely branched stems up to 2.5 m long. Because it has no true roots, coontail may float in dense mats beneath the water surface and its base is only occasionally attached to the sediment. It may also be found near the bottom in deep water, for example in creek channels.

Distribution: Coontail grows in all 50 states and Puerto Rico. It is usually found in freshwater reaches of tributaries with moderate to high nutrient concentrations and is also seen in some lower salinity tidal areas (e.g., the Middle River, the Potomac River near Alexandria, and Lake Placid on the Magothy River). It can also be found within large beds of other SAV species.

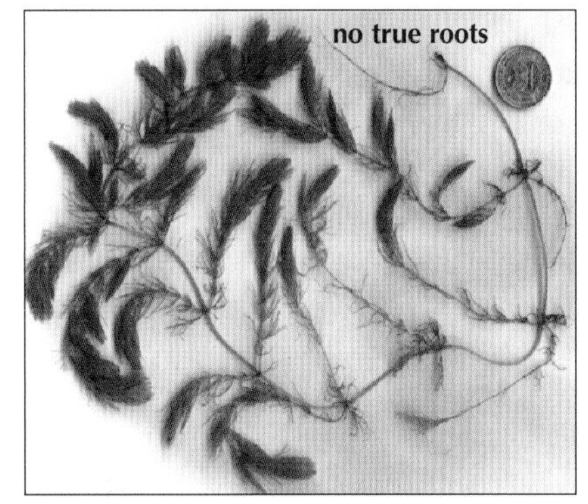

Reproduction: Asexual and sexual. Coontail reproduces asexually when stem fragments with lateral buds develop into new plants. In autumn, stem tips break off and overwinter on the bottom before sprouting in spring. Occasional sexual reproduction produces small purple flower clusters (monoecious) between July and September, followed by nutlike seeds.

Remarks: Coontail is shade tolerant and free-floating, making it less sensitive to turbid water conditions. It is an important food source for waterfowl as well as a key shade, shelter, and spawning medium for some fish.

Can be confused with: Eurasian watermilfoil (*Myriophyllum spicatum*). Eurasian watermilfoil can be distinguished from coontail by its roots and featherlike leaves, which are limp when out of water. Coontail can also be confused with parrot feather (*Myriophyllum brasiliense*). See more information on confusing species (p. 49).

SAV: COONTAIL

Best growth: 0-6.5 ppt

low to medium salinity

Other names:
hornwort

SAV: WATER STARWORT

Callitriche spp.
Family: Callichtrichaceae

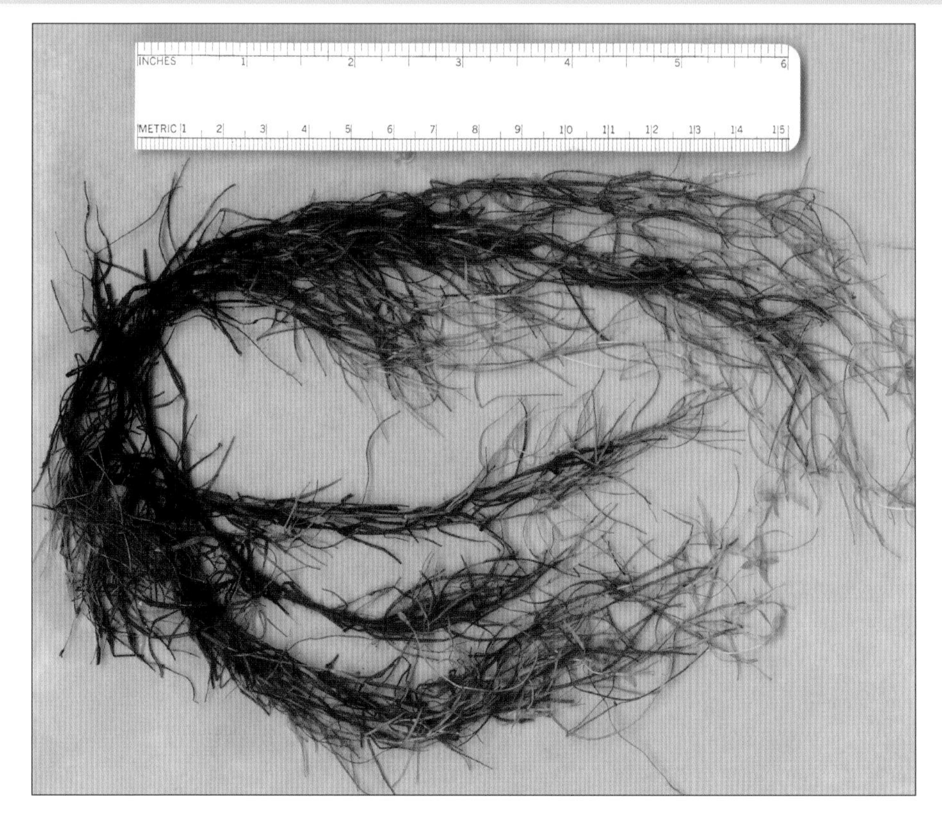

Native to Chesapeake Bay

Recognition: In North America, there are four aquatic species of water starwort that resemble each other so closely that they can only be distinguished with a hand lens or microscope when they have seeds. These species are small with very bright green leaves that are oval or obovate (egg-shaped) with the narrow end attached to the stalk. Leaf tips are obtuse (angle >90°) with two leaves growing at each joint. Leaves are 2 cm long and 3 to 8 mm wide. When water starwort grows on mud in low water conditions, its leaves are oblong. Upper leaves are obovate to oval and often float or emerge from the water.

Distribution: Water starwort grows in quiet nontidal fresh waters and occasionally in upper tidal waters. It can be found in all 48 continental states. In Chesapeake Bay, it grows in small, shallow tidal creeks and rivers.

Reproduction: Sexual and asexual. Flowering occurs July to September. Small monoecious flowers grow together in groups of 1 to 3 in the axils of leaves. The flowers, which do not have an outer

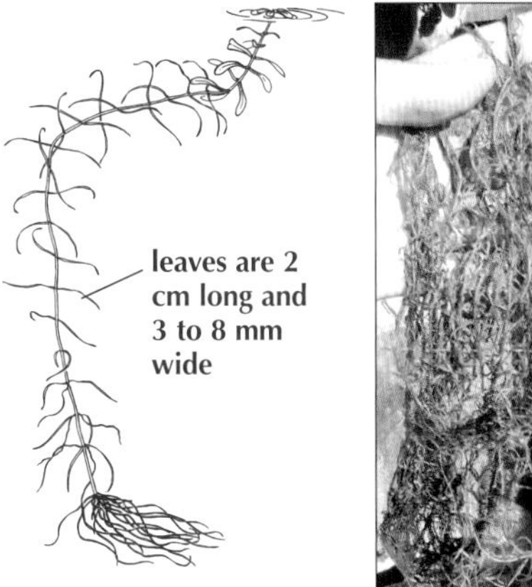

seeds

2 leaves at each joint

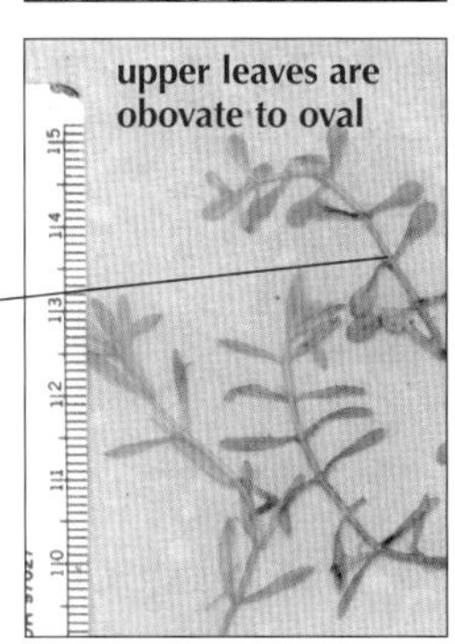

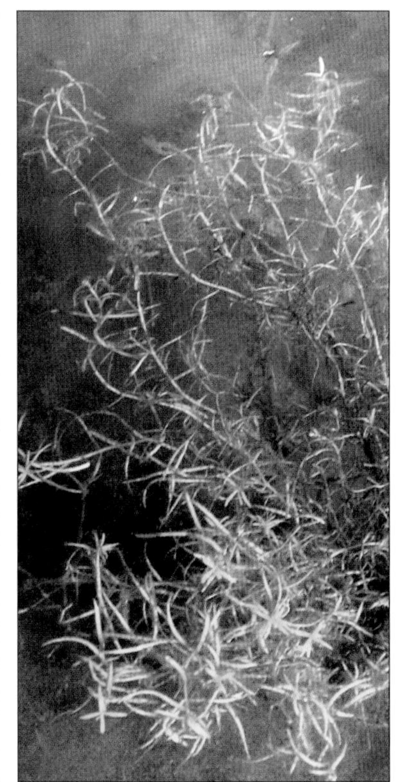

envelope (perianth), form numerous seeds closely set within the leaf axils.

Remarks: Water starwort provides forage and habitat for aquatic insects and fish. Foliage and seeds are an important food source for ducks. One species of water starwort grows on land.

Can be confused with: Waterweeds (*Elodea* spp.). Common waterweed can be distinguished from water starwort by their leaves, which grow in whorls of 3 and have a more pointed shape. Water starwort can also be confused with hydrilla (*Hydrilla verticillata*) and Brazilian waterweed (*Egeria densa*). See more information on confusing species (p. 50).

SAV: WATER STARWORT

Best growth: 0-2 ppt

low salinity

Other names:
water chickweed

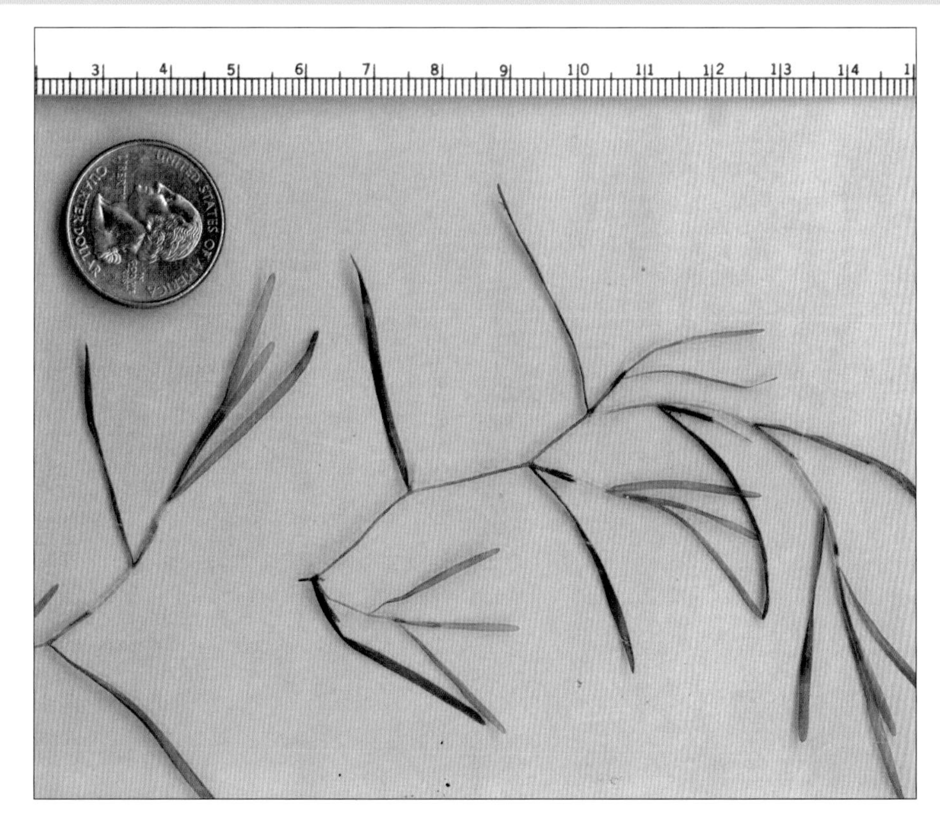

Native to Chesapeake Bay; can be invasive in ponds and reservoirs

Recognition: The leaves of slender pondweed are thin, linear, and grasslike — arranged alternately on slender, branching stems. Leaves are entire, without teeth along the edges and with a visible midrib. The leaf blades have pointed tips and can have a purplish tint. There are stipules (5-15 mm long) located below the base of the leaves, where a branching stem attaches to the main stem. The leaf base itself usually contains two small translucent glands. Slender pondweed has a root-rhizome system. In late summer, flowers occur in whorls of 3 to 5 at the end of spikes that grow between leaf bases and stems.

Distribution: Found in Chesapeake tidal tributaries, it often grows in soft, fertile mud in fresh to slightly brackish water that is quiet or gently flowing.

Reproduction: Sexual and asexual. For asexual reproduction, buds (turions) made of dense leaf clumps eventually fall off, rest during the winter, and form new plants in spring. Smooth-leaved winter buds form along branches in the axils and at the tips of

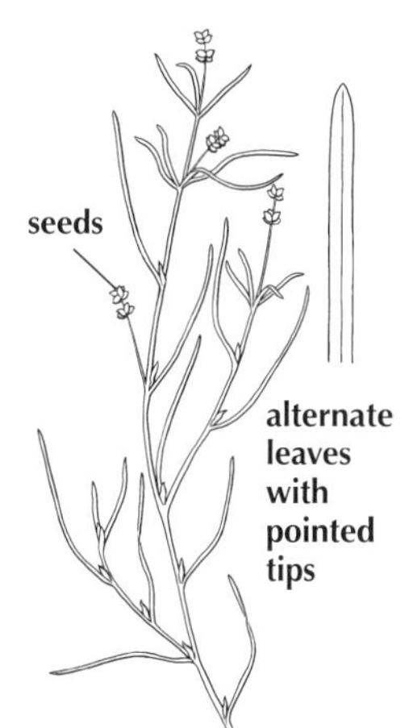

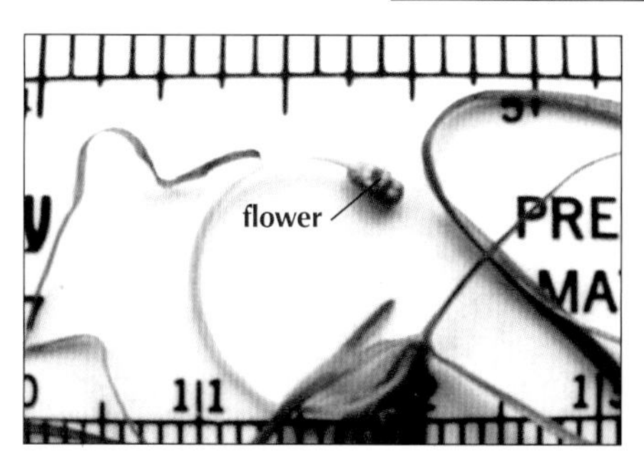

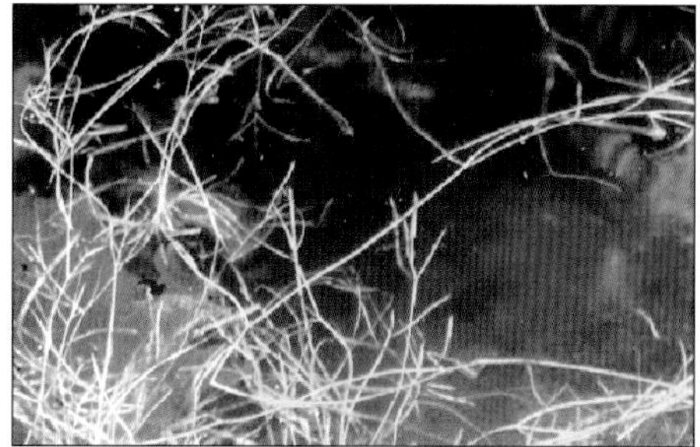

stems. Sexual reproduction occurs in the late summer with perfect flowers. Fertilization takes place underwater, producing smooth seeds with rounded backs.

Remarks: Like all other pondweeds, slender pondweed provides an important food for waterfowl. In some ponds in Maryland, slender pondweed is considered a nuisance, despite the habitat it provides for fish and invertebrates.

Can be confused with: Sago pondweed (*Stuckenia pectinata*), widgeon grass (*Ruppia maritima*), horned pondweed (*Zannichellia palustris*). Only horned pondweed usually grows in the same distribution area, but slender pondweed has been found growing with sago pondweed and widgeon grass in Grays Creek off the Magothy River. All three similar species have narrower leaves and lack the slight purplish tint of slender pondweed. See more information on confusing species (p. 51).

Best growth: 0-4 ppt

low salinity

Other names:
small pondweed

SAV: WATER STARGRASS

Heteranthera dubia
Family: Pontederiaceae

Native to Chesapeake Bay

Recognition: Grasslike leaves with no distinct midrib grow alternately on freely branching stems. The base of the leaf forms a sheath that wraps around the stem. In summer, water stargrass produces bright yellow, starlike flowers, each roughly the size of a nickel, which project above the water's surface. A land-growing form of water stargrass can be found when low water levels strand plants on shore. The land-based form has a waxy cuticle and small or leathery leaves. It produces flowers, but there is little to no branching of stems.

Distribution: Water stargrass occurs in nontidal and tidal freshwater areas of tributaries and in streams, lakes, and ponds. Water stargrass can also be common in the upper tidal Potomac River. It has also been reported in the Susquehanna Flats and the Bush, Elk, Sassafras, Middle, and Magothy rivers. It grows primarily in sediment that is clayey or chalky, but can sometimes grow in gravel streams. Water stargrass tolerates water that is moderately rich in nutrients.

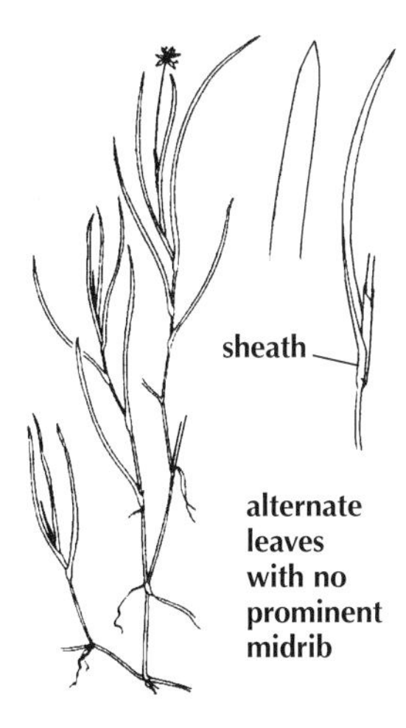

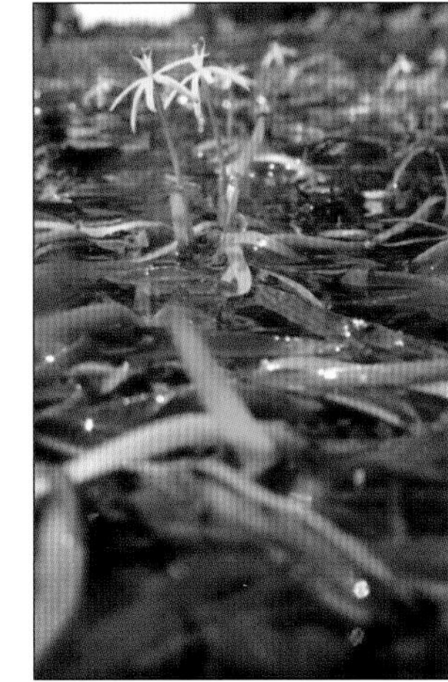

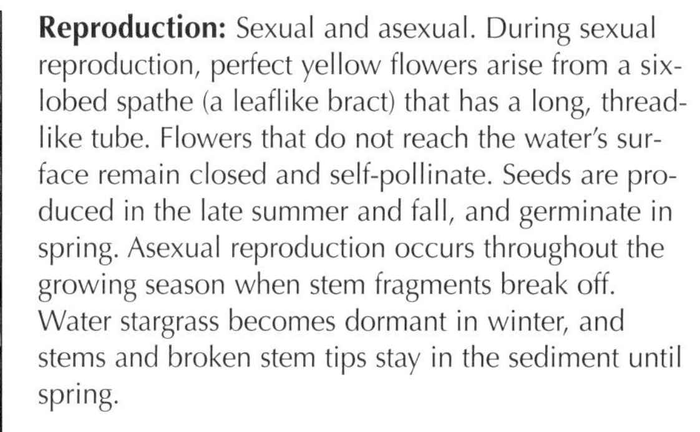

Reproduction: Sexual and asexual. During sexual reproduction, perfect yellow flowers arise from a six-lobed spathe (a leaflike bract) that has a long, thread-like tube. Flowers that do not reach the water's surface remain closed and self-pollinate. Seeds are produced in the late summer and fall, and germinate in spring. Asexual reproduction occurs throughout the growing season when stem fragments break off. Water stargrass becomes dormant in winter, and stems and broken stem tips stay in the sediment until spring.

Remarks: During the summer, the plant's bright yellow flower projects conspicuously above the water's surface.

Best growth: 0-4 ppt

low salinity

SAV: WILD CELERY

Vallisneria americana
Family: Hydrocharitaceae

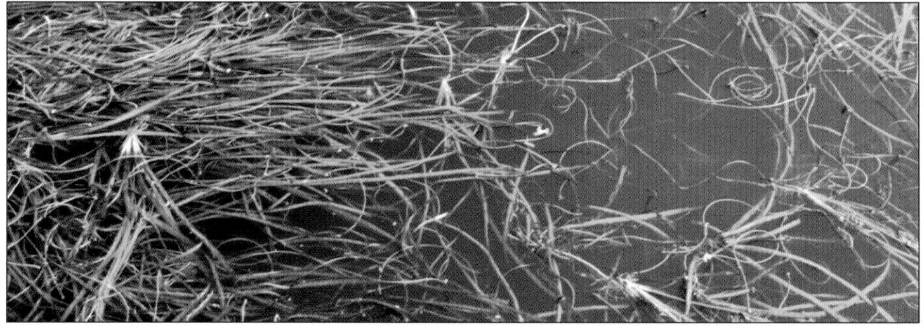

Native to Chesapeake Bay

Recognition: Long, flat, ribbonlike leaves with finely toothed edges and blunt, rounded tips arise from a cluster at the plant's base. The leaves, which have fine veins and a light green stripe down the center, grow to 1.5 m long and about 1 cm wide.

Distribution: Wild celery is found from the Atlantic coastal plain states west to Wisconsin and Minnesota. Primarily a freshwater species, it can grow occasionally in brackish waters (up to 12-15 ppt). Wild celery prefers soil that is coarse and silty to sandy. It tolerates murky waters and high nutrient levels fairly well and withstands wave action better than many other SAV species.

Reproduction: Sexual and asexual. In asexual reproduction, winter buds, or turions, form at the meristem in late summer and elongate in spring, sending a shoot to the surface from which a new plant emerges. During the growing season, each plant can send out rhizomes that grow next to the parent plant. Sexual reproduction occurs in late July to September. Wild celery is dioecious, so each plant is either male or female. Each female (pistillate) flower has three

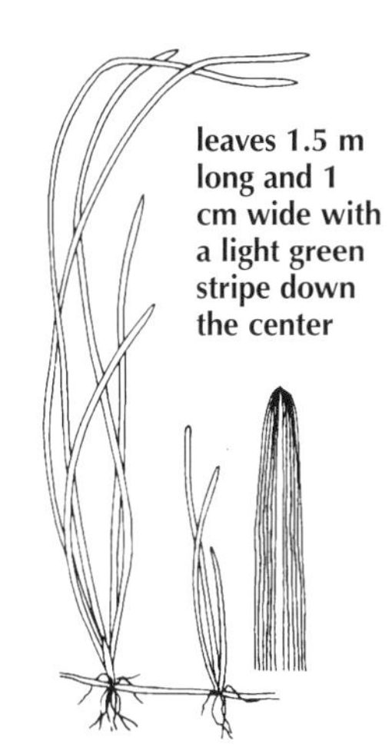

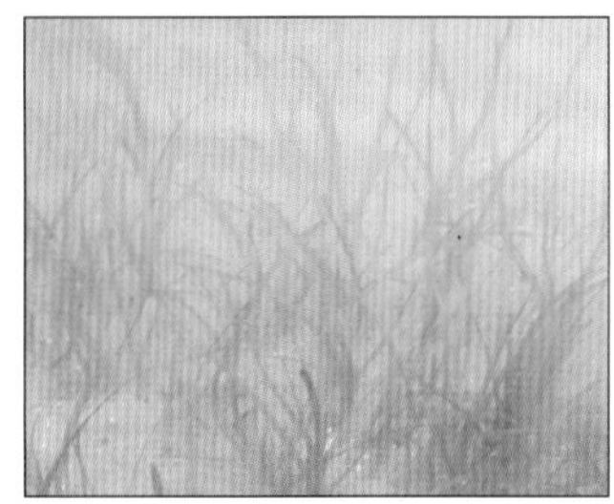

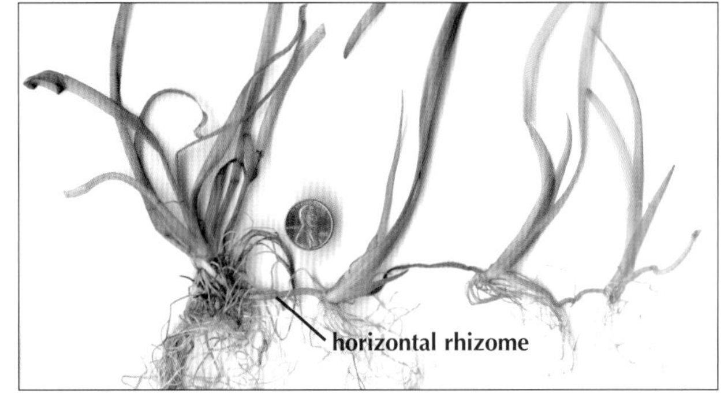

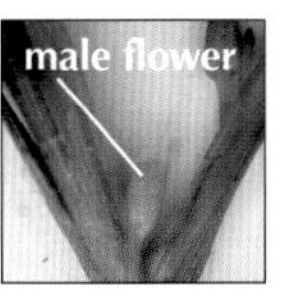

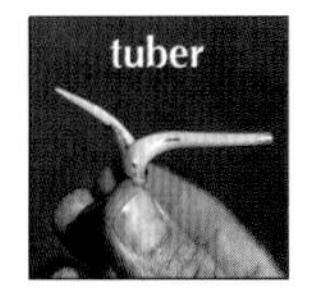

sepals and three white petals in a tubular spathe that grows to the water's surface at the end of a long stalk (peduncle). Male (staminate) flowers are crowded into an egg-shaped spathe on a short stalk near the plant's base. Eventually the spathe of male flowers breaks free, floats to the surface, and releases its pollen, fertilizing the female flowers on contact. After fertilization, the female flower's peduncle coils and the fruit develops underwater in a long, cylindrical pod containing small, dark seeds.

Remarks: Wild celery is so valuable as a source of food for waterfowl that the scientific name for the canvasback duck (*Aythya valisineria*) comes from this association. Waterfowl rely on its turions and rootstocks for food during migration and in their winter habitats. It also provides habitat for fish and invertebrates.

Can be confused with: Eelgrass (*Zostera marina*), and shoal grass (*Halodule wrightii*). Narrower leaves that lack a green stripe in the center distinguish eelgrass, also known as tape grass, from wild celery. Neither eelgrass nor shoal grass is known to co-occur with wild celery due to different salinity tolerances. See more information on confusing species (p. 52).

Best growth: 0-5 ppt

low salinity

Other names: tapegrass, freshwater eelgrass, water celery

SAV: CURLY PONDWEED

Potamogeton crispus
Family: Potamogetonaceae

Non-native to Chesapeake Bay; not invasive in tidal waters, invasive in nontidal waters

Recognition: Leaves are 3 to 10 cm long, broad, linear and finely-toothed, with curly margins. Leaves are arranged alternately or slightly opposite on flat, branched stems. Roots and rhizomes are shallow. The curly pondweed life cycle has three stages: winter form, spring/summer form, and dormant vegetative (asexual) bud. Vegetative buds sprout in the fall. The winter form of the plant has blue-green leaves that are more flattened. The spring/summer form has reddish-brown leaves that are wider and curlier.

Distribution: Curly pondweed grows in fresh nontidal to slightly brackish tidal waters. Likely introduced from Europe in the mid-1800s, today it occurs worldwide and is broadly distributed in streams, rivers, and reservoirs in the Chesapeake Bay watershed.

Reproduction: Sexual and asexual. Curly pondweed has lower horizontal stems (rhizomes) that extend to form new plants; it also develops asexually from burrlike structures near the stem tips. Curly pondweed also reproduces sexually from seeds

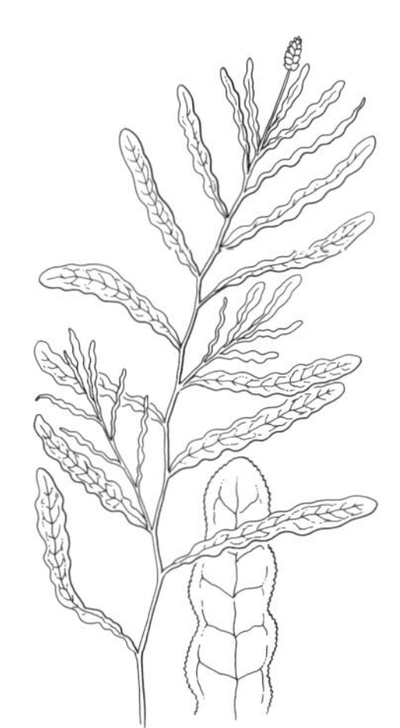

formed by perfect flowers that float at the water's surface atop spikes. Flowering occurs in late spring or early summer and the plants start dying off in midsummer after vegetative buds are produced. The buds remain dormant until fall, when the cycle is repeated.

Remarks: Curly pondweed is an introduced species widespread throughout the United States. Curly pondweed can be invasive when it occurs in lakes and reservoirs (nontidal waters).

Can be confused with: Redhead grass (*Potamogeton perfoliatus*). Only young shoots of redhead grass resemble curly pondweed.

Best growth: 0-3 ppt

low salinity

Other names:
curly leaf pondweed

Najas spp.
Family: Najadaceae

Native to Chesapeake Bay except for *N. minor*; non-invasive

Recognition: Naiads have slender branching stems. The narrow leaves with acute leaf tips (angle <90°) broaden at the base and grow opposite each other or in whorls. All leaves are less than 4 cm long. Naiads vary from 2.5 cm tufts on sandy bottoms to highly branched plants 0.6 to 0.9 m high on silty bottoms. They have small, fibrous roots without rhizomes or tubers. Naiad species are difficult to distinguish unless seeds are present, with their characteristic markings. The four species in Chesapeake Bay, listed below, look similar and are difficult to find with seeds, but some can be distinguished by their leaves.

Spiny naiad (*Najas minor*). The leaves are stiff and recurved at maturity, with teeth in the leaf margins that are visible to the naked eye. The seed coat has lengthwise ribs. (Spiny naiad is pictured at left.)

Southern naiad or bushy pondweed (*Najas guadalupensis*). The leaves are flat and straight and wider than those of both *N. minor* and slender naiad.

Slender naiad (*Najas gracillima*) and northern naiad (*Najas flexilis*) cannot be distinguished from each other easily by their leaves (see drawings below).

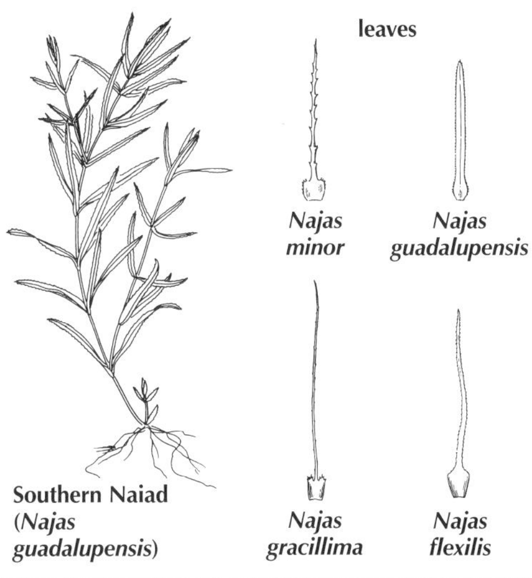

Southern Naiad (*Najas guadalupensis*)

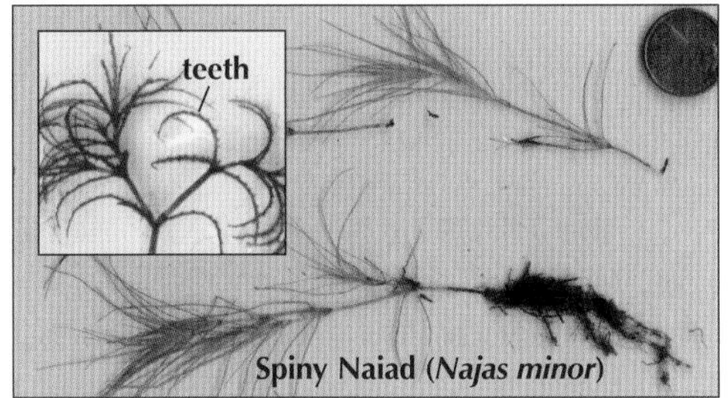

Spiny Naiad (*Najas minor*)

Southern Naiad (*Najas guadalupensis*)

Distribution: Naiads are native to Chesapeake Bay except for *N. minor*, which was introduced from Europe. Naiads grow in small freshwater streams and fresh water to brackish portions of tidal Bay tributaries. Southern naiad tolerates slightly brackish water. Naiads tolerate low light and prefer sandy bottoms.

Reproduction: Sexual only. Naiads have male and female flowers located on leaf axils. After pollination, seeds form in late summer. Seed germination and plant growth occur in spring.

Remarks: Southern naiad and northern naiad are excellent food sources for waterfowl. All parts of the plants (stems, leaves, and seeds) are eaten by a variety of waterfowl, including lesser scaup, mallards, and pintails. The other two species of naiads are not as important because they are less common; slender naiad has low nutritional value.

Can be confused with: All naiad species can be confused with each other. Look at leaves carefully to distinguish (see left). To learn more, see the online Bay Grass Key from the Maryland Department of Natural Resources: www.dnr.state.md.us/bay/sav/key.

SAV: NAIADS

Best growth: 0-5 ppt

low salinity

Other names: water nymph

SAV: HYDRILLA

Hydrilla verticillata
Family: Hydrocharitaceae

Non-native to Chesapeake Bay; invasive

Recognition: Freely branching stems have whorls of 4 to 5 simple leaves that may be linear or lance-shaped (lanceolate). Leaves have a midrib with spines and the leaf edges usually have teeth visible to the naked eye. In late summer, small white flowers float on a hypanthium, or flower receptacle, at the water's surface. Roots are adventitious; they form along nodes of rhizomes that grow horizontally along or just below the sediment. Potato-like tubers are commonly attached to the rhizomes by runners.

Distribution: Native to Africa, Australia, and parts of Asia, hydrilla was introduced to the United States during the 1960s through the aquarium trade. Today it is found in most of the southeastern United States all the way west to California. In the Chesapeake Bay region, hydrilla was first detected in 1982 in the Potomac River near Washington, D.C. By 1992, hydrilla covered 3,000 acres in the Potomac. Hydrilla is also found in the Susquehanna Flats and freshwater areas of most Bay tributaries. Although mainly a freshwater species, it has been found in salinities of 6 to 9 ppt. Hydrilla grows on silty or muddy substrates and tolerates lower light than most other SAV species.

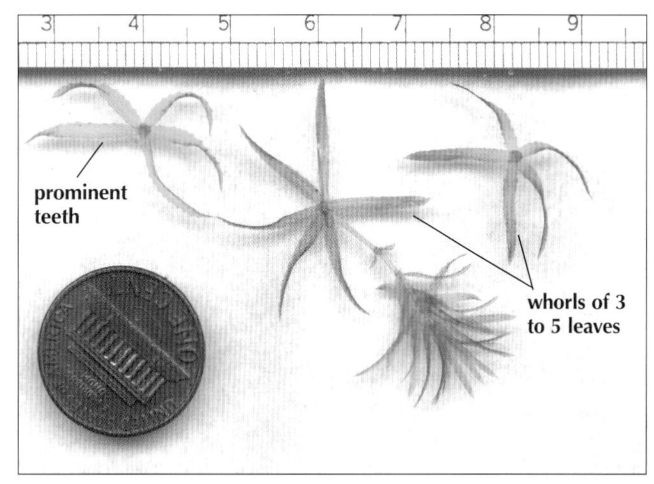

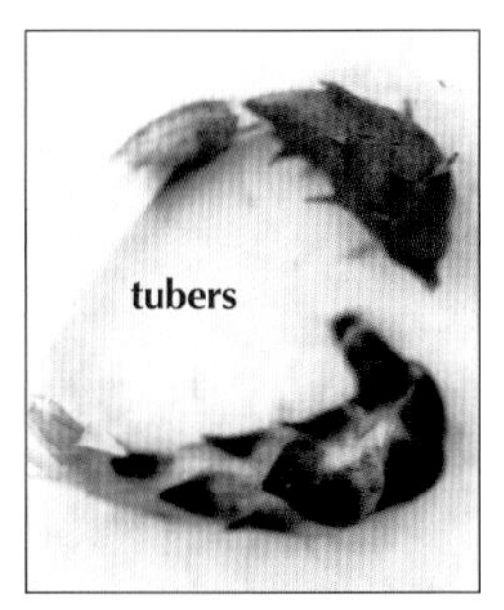

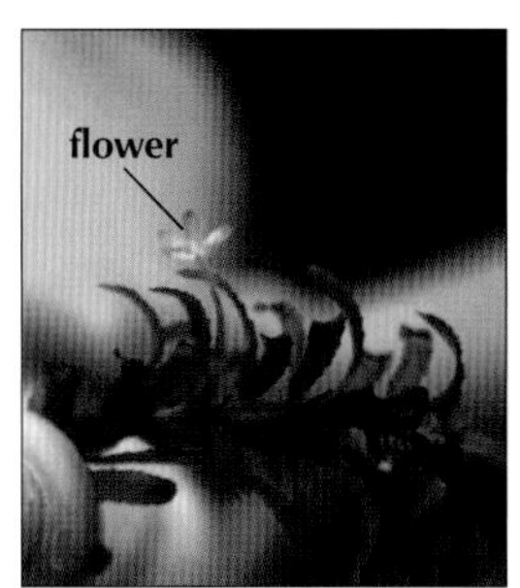

Reproduction: Sexual and asexual. The hydrilla strain in Chesapeake Bay bears both male and female flowers on one plant (monoecious), which occur together near the stem tips. The white female flowers stay attached to the plant and float on the water's surface; the male flowers detach from the stem tips and their pollen must settle directly on the female flower for sexual reproduction to occur. Seed set has a success rate of about 50 percent. Asexual reproduction is much more successful and occurs from plant fragments, tubers, rootstock, and turions (resting plant buds that form in leaf axils or stem tips). Tubers and turions can overwinter and are the major form of reproduction during the late summer, when dense hydrilla beds die off.

Remarks: Hydrilla is often considered a nuisance when it forms dense beds that interfere with recreational use of waterways. On the other hand, hydrilla is an excellent food source for waterfowl and habitat for fish. In spring, largemouth bass often frequent young hydrilla beds just reaching the water's surface.

Can be confused with: Waterweeds (*Elodea* spp.), Brazilian waterweed (*Egeria densa*), water starwort (*Callitriche* spp.). Both hydrilla and waterweeds may be called water thyme. See confusing species (p. 50).

SAV: HYDRILLA

Best growth: 0-5 ppt

low salinity

Other names: water thyme, Florida elodea

Native to Chesapeake Bay

Recognition: Two species of *Elodea* occur in Chesapeake tidal waters: Common (or Canadian) waterweed, *Elodea canadensis*, and Nuttall's waterweed, *E. nuttallii*. Unless otherwise noted, text and illustrations in this guide refer to both species, since they are difficult to distinguish except by flowers, which grow infrequently. *Elodea* has slender, branching stems and a weak, threadlike root system. In lower light conditions, plants have more space between leaf whorls and look less robust. Leaves are linear to oval with minutely toothed margins (teeth may only be visible with magnification) and blunt tips, though in general *Elodea* leaves can vary greatly in width, size, and bunching. Leaves of Nuttall's tend to be narrower (0.7-1.8 mm wide) and more linear (less oblong) than leaves of common (1.1-5 mm wide). Leaves of both species lack leaf stalks and occur in whorls of 3 at stem nodes, becoming more crowded toward the stem tips.

Distribution: Although primarily a freshwater genus, *Elodea* occasionally grows in brackish Chesapeake tributaries, often in slow-moving and/or calcium-rich waters. Nuttall's may have a higher salinity tolerance

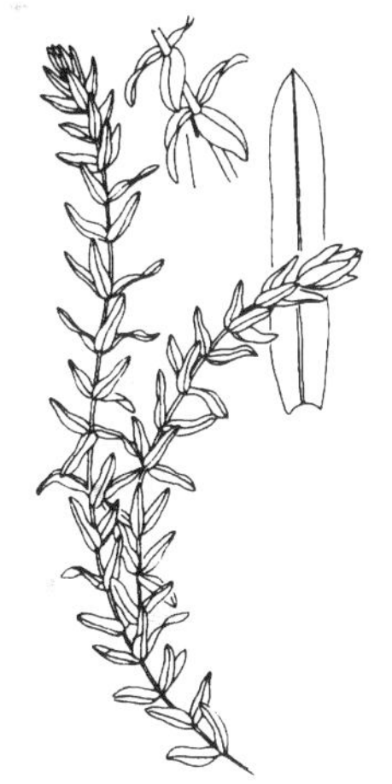

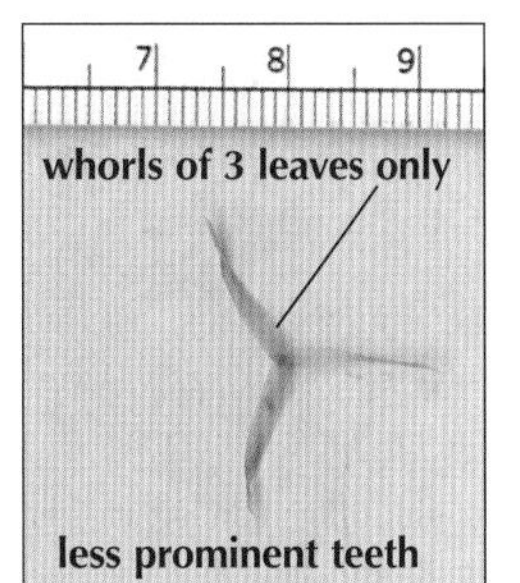

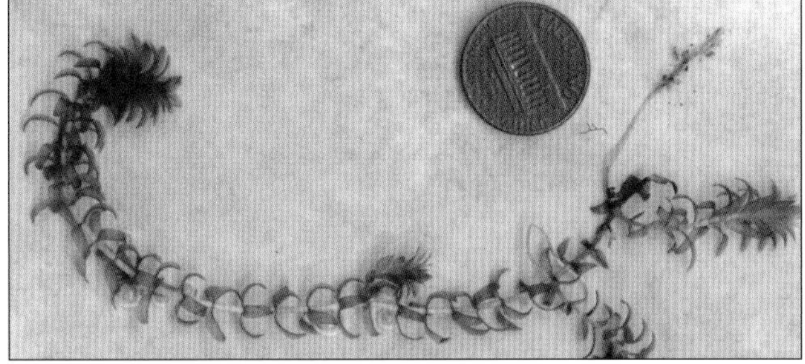

than common, but their salinity tolerances and distribution in Chesapeake Bay are not well documented. Both species are found throughout temperate North America; a third species, *E. bifoliata*, occurs only in the Western U.S.

Reproduction: Asexual and sexual. *Elodea* typically reproduces asexually through vegetative propagation. *Elodea* is dioecious: separate plants have either male or female flowers. Sexual reproduction is rare but when it occurs, plants produce cylindrical fruit capsules.

Remarks: Although *Elodea* can be invasive where it has been introduced, especially in Europe (both species) and in Australia and New Zealand (common only), it forms an important part of tidal and nontidal freshwater ecosystems and provides good habitat for many aquatic invertebrates and cover for young fish and amphibians. Waterfowl, especially ducks, as well as beaver and muskrat eat this plant.

Can be confused with: Hydrilla (*Hydrilla verticillata*), water starwort (*Callitriche* spp.), Brazilian waterweed (*Egeria densa*). See more information on confusing species (p. 50).

Best growth: 0-4 ppt

low salinity

Other names: Canadian waterweed, water thyme

SAV: EURASIAN WATERMILFOIL

Myriophyllum spicatum
Family: Haloragaceae

Non-native to Chesapeake Bay; invasive

Recognition: This mostly submerged plant grows to 2.5 m tall. It has long, underwater stems that branch and produce featherlike (pinnate) leaves that grow in whorls of 4 or 5. These compound leaves are 0.8 to 4.5 cm long with 9 to 13 hairlike segments per side. When removed from water, these delicate leaves compress and lose their shape. Lower portions of the stems may be leafless, and stems are stout and tough, especially on older plants. During late summer Eurasian watermilfoil may grow reddish flower spikes on stem tips that protrude above the water's surface.

Distribution: Accidentally introduced from Europe and Asia, Eurasian watermilfoil is now found throughout the United States. Explosive growth of Eurasian watermilfoil during the late 1950s covered large areas of Chesapeake Bay and its tidal tributaries until it died back and stabilized by 1970. Eurasian watermilfoil inhabits nontidal fresh to moderately brackish tidal water. It often grows in sandy mud in slow-moving streams or protected waters and does not tolerate strong currents or waves. It is often the first species to appear in the spring in tidal tributaries where water quality has degraded.

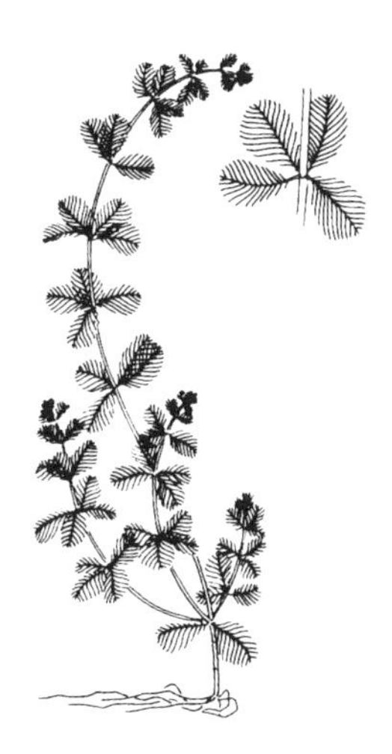

flower
spike

Reproduction: Sexual and asexual. Eurasian watermilfoil reproduces most commonly by asexual means, from stem fragments. Although it is monoecious, with male and female flowers on the same plant, it does not self-pollinate because the female (pistillate) flowers mature before the male (staminate) flowers. Pollination through the air produces nutlike fruits that sink to the bottom and stay viable for years.

Remarks: Although not a great food source for waterfowl, the plant provides excellent cover for young fish, crabs, and invertebrates. Fishermen recognize watermilfoil beds as good places to catch largemouth bass, which lie in ambush among the stems.

Can Be Confused with: Parrot feather (*Myriophyllum brasiliense*), coontail (*Ceratophyllum demersum*). See more information on confusing species (p. 49).

SAV: REDHEAD GRASS

Potamogeton perfoliatus
Family: Potamogetonaceae

Native to Chesapeake Bay

Recognition: Redhead grass is one of the most easily recognizable SAV species in Chesapeake Bay because of the distinctive way its oval leaves wrap around the stem. In relatively shallow water, plants have thicker, darker green foliage than they do in deeper water. Leaves of redhead grass are flat and oval-shaped, 1 to 7 cm long and 1 to 4 cm wide. Leaves have parallel veins and edges that curl slightly, and they grow in an alternate or slightly opposite pattern along slender, straight stems. Branching is more developed in the upper part of the plant. Redhead grass has an extensive root and rhizome system.

Distribution: Redhead grass has been documented in North America from Labrador south to Florida and west to Ohio as well as in Central America (Guatemala), Eurasia, Africa, and Australia. Redhead grass grows in many Maryland rivers. It has not been documented in Virginia waters recently but could potentially establish populations under suitable conditions. Redhead grass is typically found in fresh to moderately brackish and alkaline waters. It grows best on firm, muddy substrate and in quiet water with slow-moving currents.

Reproduction: Sexual and asexual. Sexual reproduction takes place in early to mid-summer, when spikes of tiny perfect flowers emerge from leaf axils on the ends of plant stems and extend above the water's surface. The wind carries the pollen. As fruits mature they sink and release seeds. Attempts to propagate plants from seed have been relatively unsuccessful, but rooted cuttings have been transplanted successfully. Redhead grass reproduces asexually by resting buds that develop serially at the end of a growing season from the apex of rhizomes. These buds produce the next year's spring shoots.

Remarks: Redhead grass, probably named for the redhead ducks that often eat it, is an excellent food source for waterfowl. The wide, horizontal leaves of redhead grass may be more susceptible than those of other SAV to covering by tiny plants (epiphytes, or epiphytic growth) that use them for mechanical support.

Can be confused with: Curly pondweed (*Potamogeton crispus*). Only the young shoots of redhead grass resemble curly pondweed.

SAV: REDHEAD GRASS

Best growth: 5-10 ppt

medium salinity

Other names: tea leaf, clasping leaf pondweed

SAV: HORNED PONDWEED

Zannichellia palustris
Family: Zannichelliaceae

Native to Chesapeake Bay

Recognition: Long, linear, threadlike leaves grow opposite or in whorls on slender, branching stems. Leaf tips gradually taper to a point, and a thin sheath (stipule) covers the basal parts of leaves. The plant has tendril-like roots and slender rhizomes. Its seeds, which form in the leaf axils, have a distinctive horn shape and grow in groups of 2 to 4. Two forms of horned pondweed are found in Chesapeake Bay: (1) upright with free-floating branches or (2) creeping with stem node roots that anchor the plant in areas with high wave activity. The creeping form is also common in winter. Horned pondweed is usually the first SAV to appear in spring. It declines in June or early July because it cannot tolerate high water temperatures, producing floating mats of decaying plants.

Distribution: Horned pondweed is found in every state in the continental United States, as well as in Europe and South America. It is widely distributed in Chesapeake Bay and grows in fresh to moderately brackish waters in both muddy and sandy sediments. Although horned pondweed is generally a shallow water plant, it can grow in depths of up to 5 m.

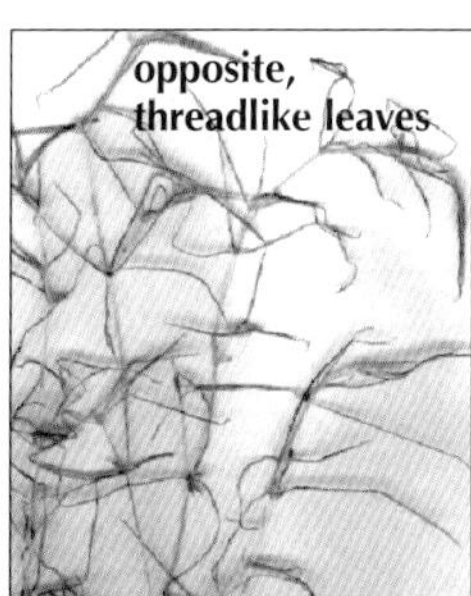

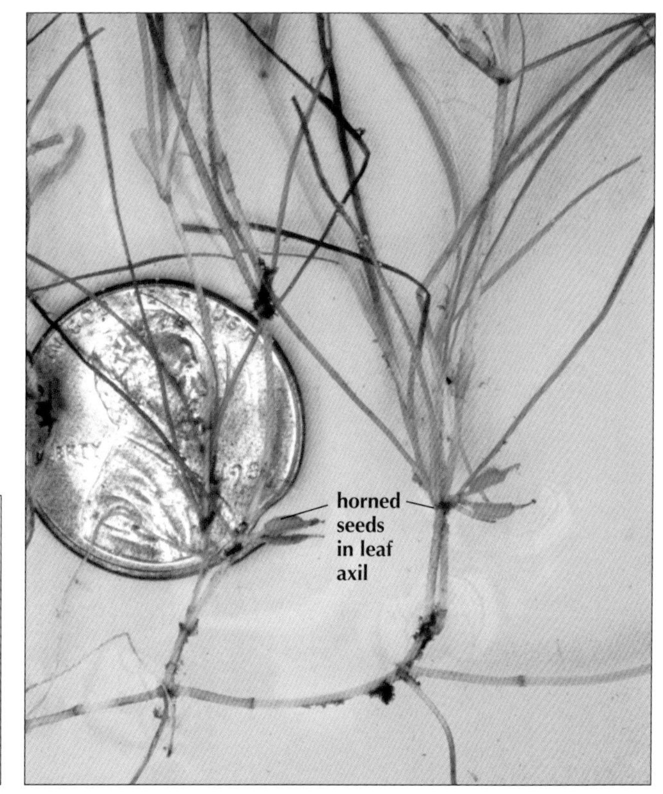

Reproduction: Sexual only. The hornlike, slightly curved seeds, which form in early to late spring, are the primary means of reproduction. Flowers are imperfect (bisexual) and monoecious. The seeds usually germinate in the same year they are set. By June, as water temperatures warm, the plants release their seeds and die back. Fruit development is also quite rapid. A second growth cycle may occur in fall, and the plant can grow over the winter in some areas.

Remarks: Both seeds and vegetative parts of this plant are an important food source for waterfowl. The creeping form of horned pondweed is found in areas of moderate wave energy, and its adventitious roots help stabilize the sediment and resist dislodgement.

Can be confused with: Sago pondweed (*Stuckenia pectinata*), widgeon grass (*Ruppia maritima*), slender pondweed (*Potamogeton pusillus*). See more information on confusing species (p. 51).

Best growth: 0-11 ppt

low to medium salinity

SAV: SAGO PONDWEED

Stuckenia pectinata
(formerly *Potamogeton pectinatus*)
Family: Potamogetonaceae

Native to Chesapeake Bay

Recognition: Sago pondweed's threadlike leaves are 3 to 10 cm long and 0.5 to 2 mm wide. The leaves taper to a point and are arranged in an alternate pattern. The sheath of leaves at the base sometimes has a whitish pointed tip or bayonet that helps identify the plants when they are not in flower. The slender stems are abundantly branched, and bushy leaf clusters fan out at the water's surface. Seeds form in clusters at the tips of the stems. Roots are long and straight with slender rhizomes.

Distribution: Found throughout the United States, South America, Europe, Africa, and Japan, sago pondweed is widespread in Chesapeake Bay, where it grows in fresh nontidal to moderately brackish waters. In mesohaline rivers where it grows near its lower salinity limit, such as the Magothy River, it tends to be less abundant in high rainfall/low salinity years. It can tolerate high alkalinity and grows on sediments that are silty or muddy. Because of its long rhizomes and runners, sago pondweed tolerates strong currents and wave action better than most SAV species.

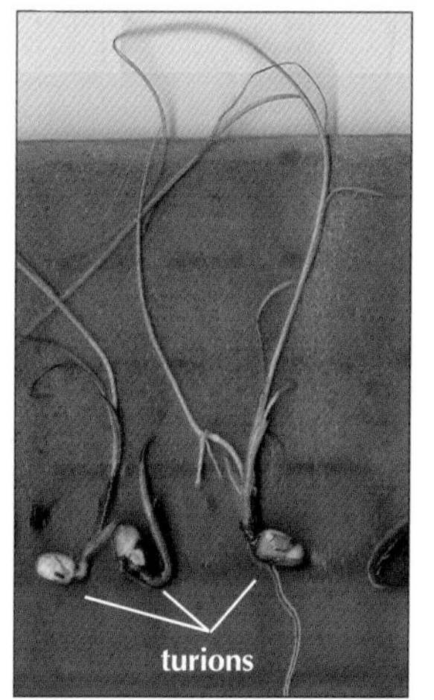

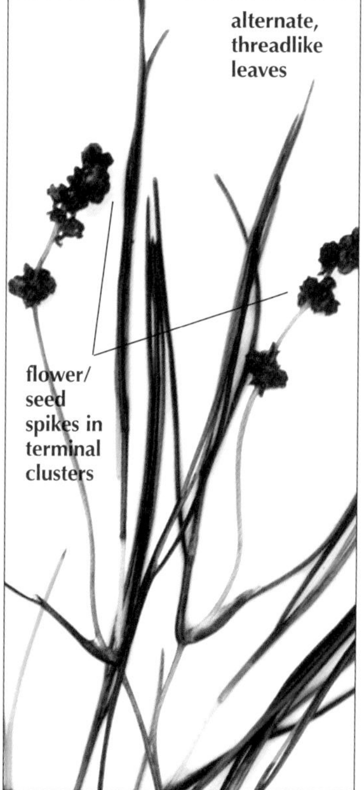

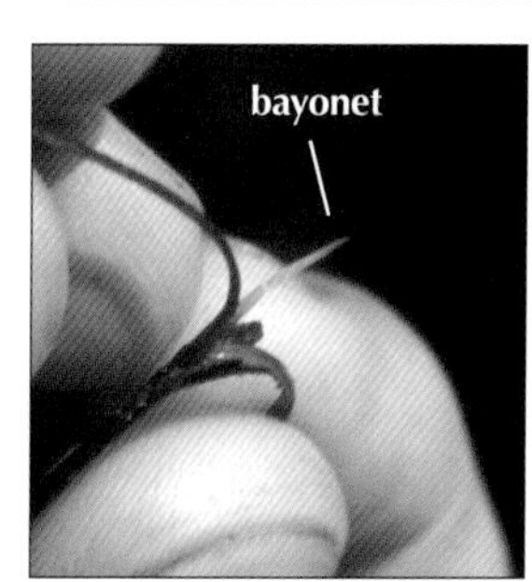

Reproduction: Sexual and asexual. In early summer, perfect flowers that look like beads grow along a thin terminal spike. The flowers release pollen that floats on the water's surface. After fertilization, seeds develop on the shaft of the spike and remain until fall, when they spread into the water. The germination rate is low, however. More commonly, asexual reproduction occurs by two kinds of starchy tubers: (1) tubers (called turions) produced at the ends of underground rhizomes and runners or (2) tubers that form in leaf axils and at the end of leaf shoots and later drop off and sink to the bottom. Both kinds of tubers grow singly or in pairs and can form plants later in spring.

Remarks: Sago pondweed is one of the most valuable food sources for waterfowl in North America. Many species of ducks, geese, swans, and marsh and shorebirds eat its seeds and tubers, which are full of nutrients, as well as its leaves, stems, and roots. It grows well in tanks from tubers or using micropropagation.

Can be confused with: Horned pondweed (*Zannichellia palustris*), widgeon grass (*Ruppia maritima*), slender pondweed (*Potamogeton pusillus*). See more information on confusing species (p. 51).

SAV: SAGO PONDWEED

Best growth: 0-12 ppt

low to medium salinity

Other names:
fennel pondweed

SAV: WIDGEON GRASS

Ruppia maritima
Family: Ruppiaceae

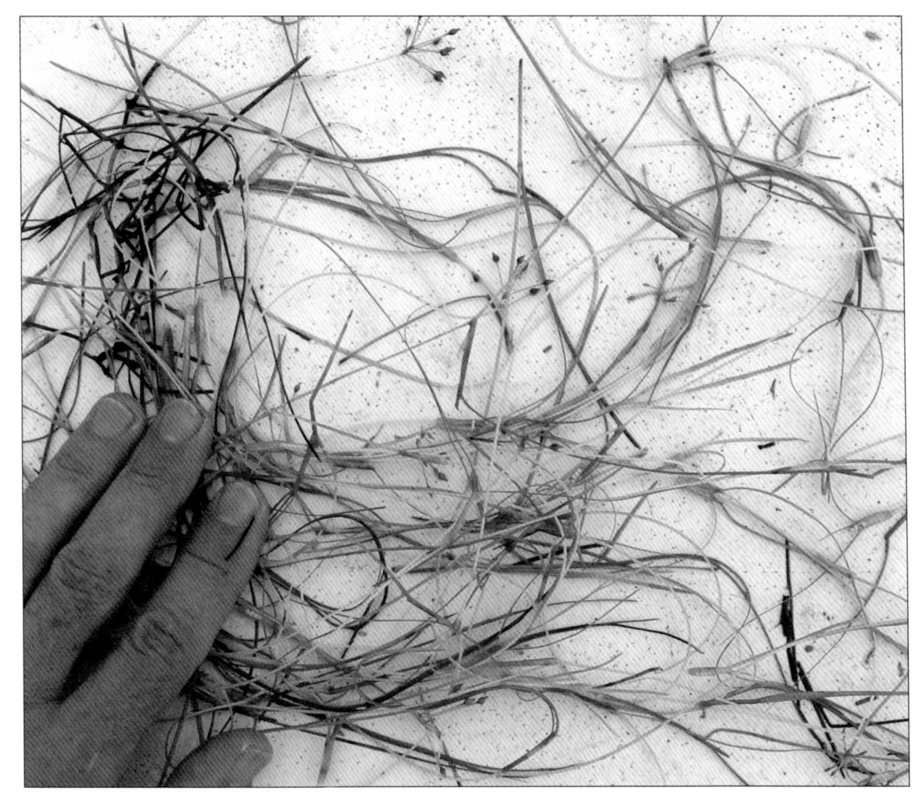

Native to Chesapeake Bay

Recognition: Straight, threadlike leaves are 3 to 10 cm long and 0.5 mm or less wide. Leaves have a sheath at their base and rounded tips, and grow alternately along slender, branching stems. Widgeon grass has an extensive root system of branched, creeping rhizomes that lack tubers. Widgeon grass grows in two forms in Chesapeake Bay: (1) upright and highly branched in appearance, with flowers standing several feet tall, and (2) shorter and creeping, with basal leaves that can be present at any time of year, although this form is the only one that may overwinter. When in seed, widgeon grass has single seed pods that form at the end of fan-shaped clusters of short stalks.

Distribution: Widgeon grass tolerates a wide range of salinity, from the slightly brackish upper and mid-Bay tributaries to near seawater salinity in the lower Bay and high salinity in salt pannes — small ponds formed in depressions in salt marshes. Widgeon grass has also been found in the freshwater parts of some estuaries and in nontidal waters. In more saline lower Bay areas (mainly in Virginia), widgeon grass and eelgrass are the dominant SAV species. Widgeon grass is

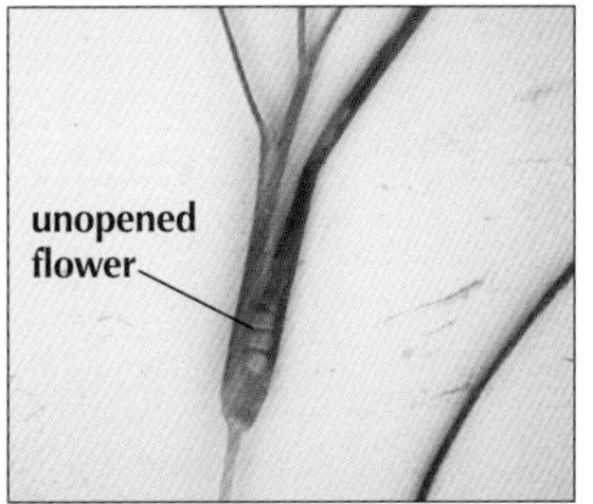

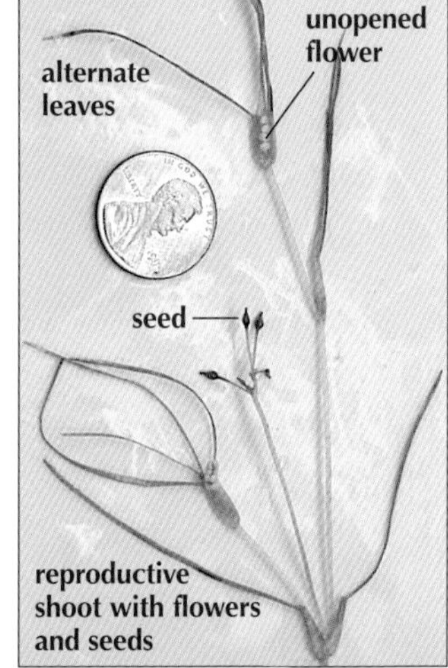

short, creeping form

most common in shallow areas with sandy bottoms, occasionally growing on soft, muddy sediments. High wave action can damage the slender stems and leaves.

Reproduction: Sexual and asexual. Between late spring and late summer, two perfect flowers emerge enclosed in a basal sheath of leaves. The flowers extend towards the water's surface on a stalk; pollen released from the stamen floats on the water until it contacts one of the extended pistils. Fertilized flowers produce four black, oval-shaped fruits with pointed tips. The individual fruits are at the end of stalks that grow in clusters of eight. Widgeon grass reproduces asexually when new stems emerge from the root-rhizome system.

Remarks: Widgeon grass is a valuable food for waterfowl. In saltier water, widgeon grass often grows beside eelgrass, with eelgrass more common in deeper water. Widgeon grass can also grow in ditches along roadsides and in those draining agricultural fields.

Can be confused with: Horned pondweed (*Zannichellia palustris*), sago pondweed (*Stuckenia pectinata*), slender pondweed (*Potamogeton pusillus*). See more information on confusing species (p. 51).

SAV: WIDGEON GRASS

Best growth: 5-15 ppt

medium salinity

Other names:
ditch grass

SAV: EELGRASS

Zostera marina
Family: Zosteraceae

Native to Chesapeake Bay

Recognition: Eelgrass has a thick, creeping rhizome, 2 to 5 cm long, with many roots and nodes spaced 1 to 3.5 cm apart. Alternate, ribbonlike leaves with rounded tips arise from these nodes and grow to 1.2 m long and 2 to 12 mm wide. At the base of each leaf is a tube-shaped membranous sheath that measures 5 to 20 cm long and wider than the leaf itself. When eelgrass grows in shallow, sandy, exposed substrates the leaves are relatively small and narrow. Plants in deep and muddy areas have longer, wider leaves.

Distribution: Eelgrass grows along both coasts of the United States and worldwide. The North Carolina coast marks the southern limit of its distribution on the East Coast and Labrador, Canada is its northern limit. Eelgrass is one of the most abundant and persistent SAV species in high-salinity portions of Chesapeake Bay and its lower tributaries; most of it is found from the Honga River south to the mouth of the Bay. Along with widgeon grass, eelgrass is the dominant SAV species in the lower Bay (mainly in Virginia) and the coastal bays. Small, isolated populations of eelgrass have also been found in Maryland's Eastern Bay.

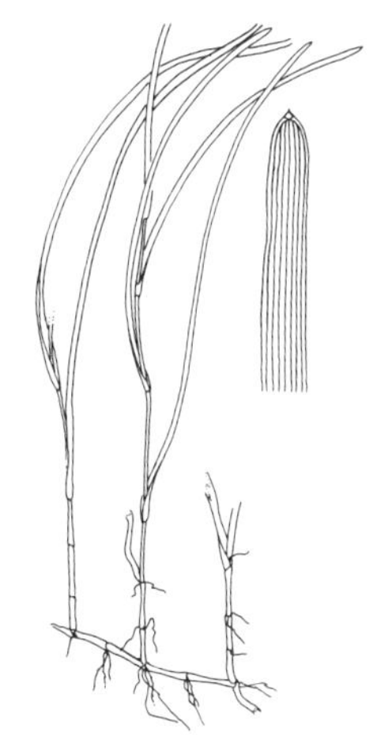

Reproduction: Sexual and asexual. Plants flower in May and June. Each plant has both male and female flowers (monoecious), which mature at different times on the same plant to prevent self-fertilization. Flowers, fertilized by drifting pollen, develop into seed-bearing generative shoots that eventually break off, float to the surface, and release their seeds as they drift. For asexual reproduction, the rhizome grows and elongates and turions, or winter buds, form.

Remarks: Eelgrass is one of three true "sea grasses" covered in this guide; the other two are widgeon grass and shoal grass. Eelgrass is important habitat for blue crabs, which use the beds for protective cover as juveniles and during shedding and mating. It also provides important habitat for the bay scallop, seahorse, pipefish, and speckled sea trout. Eelgrass is a major food source for brant (which almost disappeared from Chesapeake Bay when "wasting disease" decimated eelgrass in the 1930s), Canada geese, widgeon, redhead and black ducks, and green sea turtles.

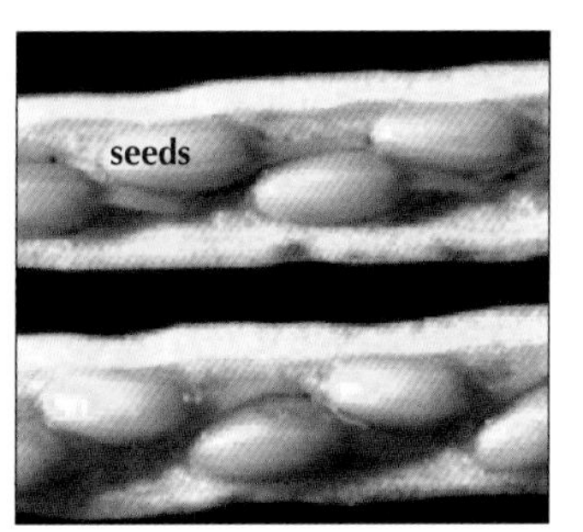

seed pods after releasing seeds

Can be confused with: Wild celery (*Vallisneria americana*), shoal grass (*Halodule wrightii*). Wild celery is also called tapegrass. See more information on confusing species (p. 52).

SAV: EELGRASS

Best growth: 15-30 ppt

medium to high salinity

Other names:
tapegrass

SAV: SHOAL GRASS

Halodule wrightii
Family: Cymodoceanceae

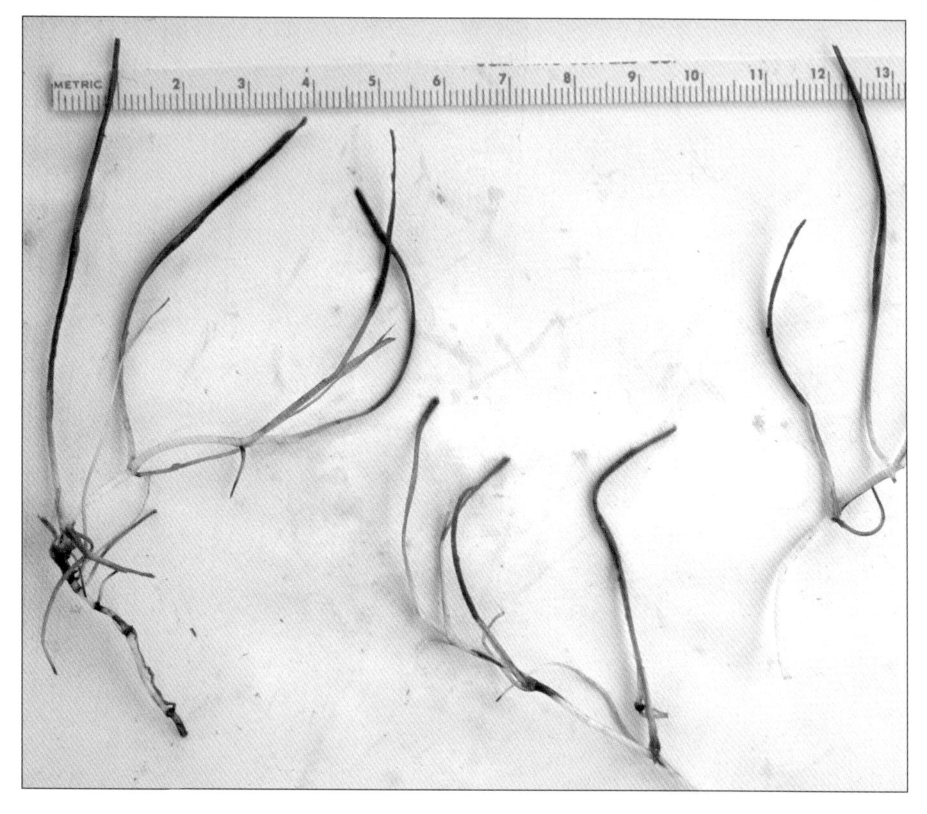

Not found in Chesapeake Bay; found as far north as North Carolina coast

Recognition: Shoal grass has green leaves that grow in a cluster from shoots along a rhizome, with distinct nodes spaced 0.5 to 4 cm apart and a prominent midvein. On average, 2 to 5 roots and one leafy shoot emerge from each node. The leaf blades are long and thin and can range from 3.5 to 32 cm long and 0.3 to 2.2 mm wide. The leaf tip is concave and bidentate, sometimes with a small central point. Its flowers are inflorescent, all growing from a single axis. Shoal grass produces egg-shaped fruits with dimensions ranging from 1.5 to 2 mm by 1.2 to 1.8 mm.

Distribution: Shoal grass can be found south from North Carolina along the Atlantic and Gulf coasts, with a gap in its distribution in South Carolina, Georgia, and northern Florida. It is found in the Caribbean to temperate South America, northwest Africa, and possibly in the Indian Ocean and the Pacific coast of Mexico.

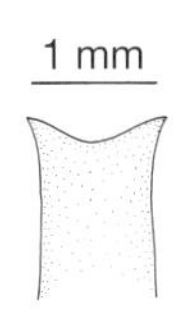

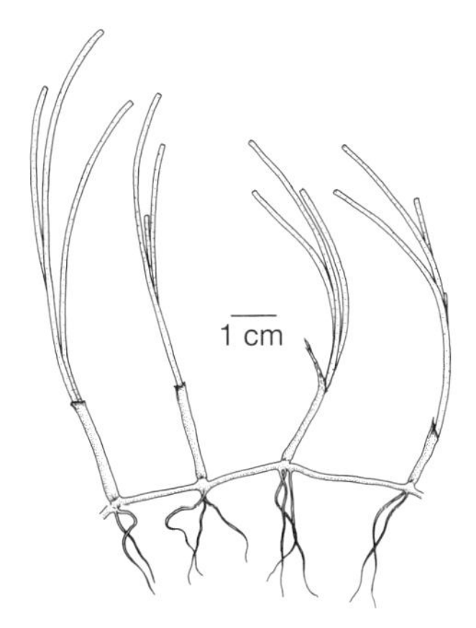

Reproduction: Sexual and asexual. Shoal grass is dioecious, but flowering is rare. Shoal grass can rapidly and densely recolonize bare areas in warm months. New shoot production probably occurs through rhizome elongation.

Remarks: Shoal grass can be found on a variety of substrate types, ranging from silty mud to coarse sand. It is a food source for sea turtles, sea urchins, and several fish species, and shelters a diverse assemblage of invertebrates such as crabs and amphipods. As many as 41 different epiphytic plants grow on shoal grass and provide an important food source for certain gastropod grazers.

Can be confused with: Eelgrass (*Zostera marina*), widgeon grass (*Ruppia maritima*). See more information on confusing species (p. 52).

Best growth: 20-35 ppt

high salinity

Other names: shoal weed

Coontail

Identifying Confusing SAV Species

On the pages that follow, you will find additional information about four groups of SAV species in Chesapeake Bay and Mid-Atlantic coastal waters that share similar characteristics and can sometimes be confused with each other. These pages include photographs and descriptions of distinguishing characteristics for each plant — such as leaf growth pattern, leaf shape, coloration, or preferred salinity range. When used along with the multichotomous key (pp. 8-15), these pages should help you differentiate the following four groups of confusing species.

Eurasian watermilfoil (*Myriophyllum spicatum*)
Coontail (*Ceratophyllum demersum*)
Parrot feather (*Myriophyllum brasiliense*)

Waterweeds (*Elodea* spp.)
Hydrilla (*Hydrilla verticillata*)
Water starwort (*Callitriche* spp.)
Brazilian waterweed (*Egeria densa*)

Eurasian Watermilfoil

Slender pondweed (*Potamogeton pusillus*)
Sago pondweed (*Stuckenia pectinata*)
Horned pondweed (*Zannichellia palustris*)
Widgeon grass (*Ruppia maritima*)

Eelgrass (*Zostera marina*)
Shoal grass (*Halodule wrightii*)
Wild celery (*Vallisneria americana*)

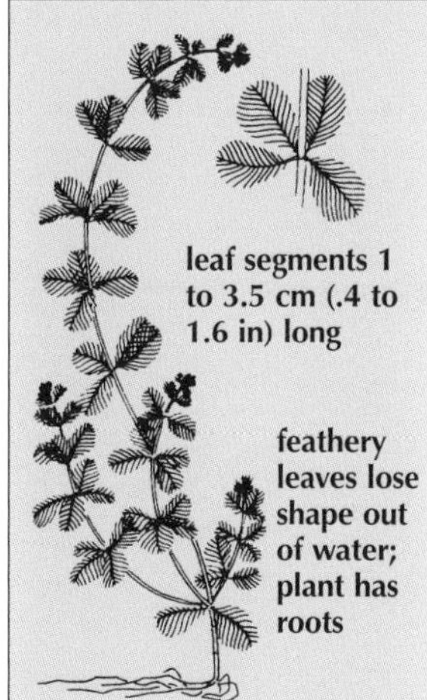

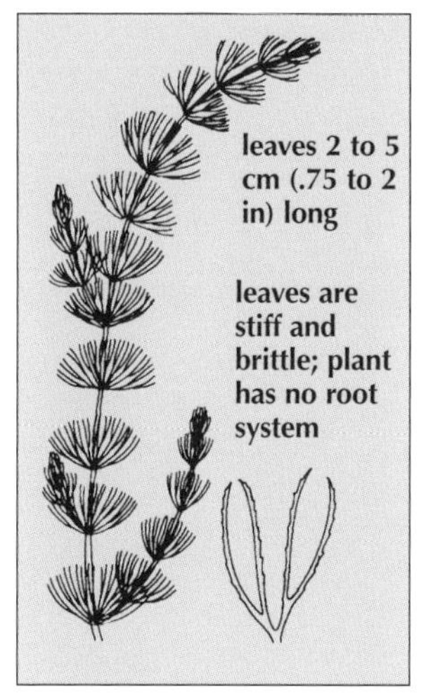

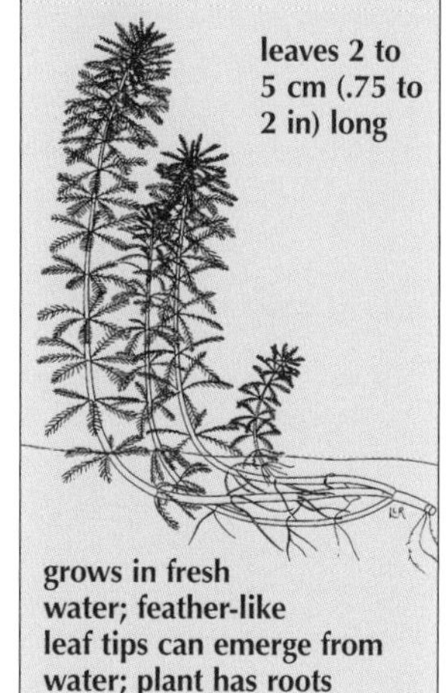

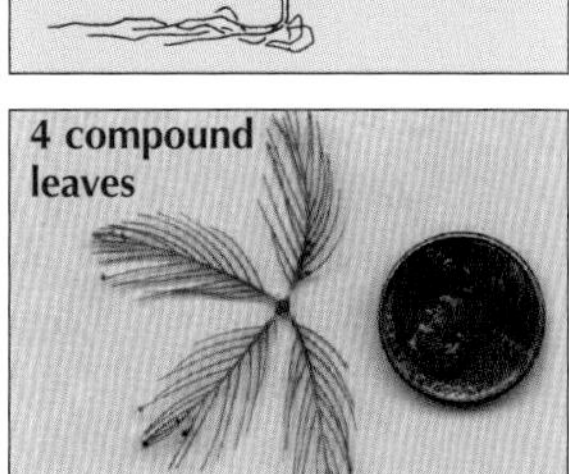

Eurasian Watermilfoil
Myriophyllum spicatum, p. 34

Coontail
Ceratophyllum demersum, p. 16

Parrot Feather
Myriophyllum brasiliense

Eurasian watermilfoil (*Myriophyllum spicatum*) can be confused with coontail (*Ceratophyllum demersum*). It can be distinguished from coontail by its root system (coontail lacks one), feathery, limp leaves, and whorls of 4 to 5 compound leaves at stem nodes instead of coontail's 9 to 10 simple leaves.

Eurasian watermilfoil can also be confused with parrot feather (*Myriophyllum brasiliense*), a species not included in this guide. At present, parrot feather has not been found in Chesapeake tidal waters, but since it is sold in aquarium and aquatic pond stores it does have the potential to invade them, and it is also invasive in nontidal ponds. Parrot feather can be distinguished from Eurasian watermilfoil by its whorls of 5 to 6 compound leaves, but like milfoil it has roots and loses shape out of water. It is limited to fresh water and can have emergent gray-green leaf tips.

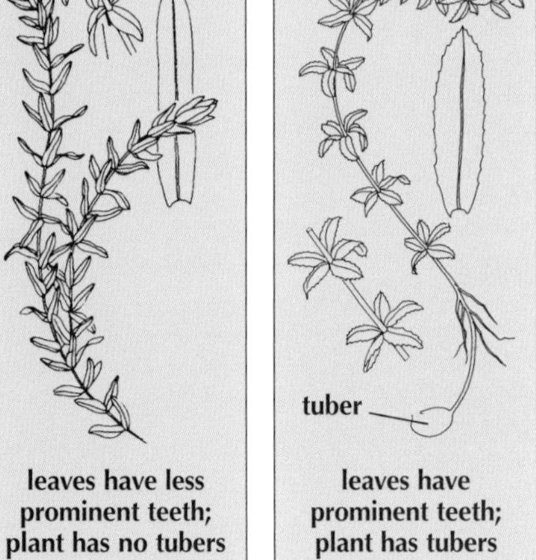

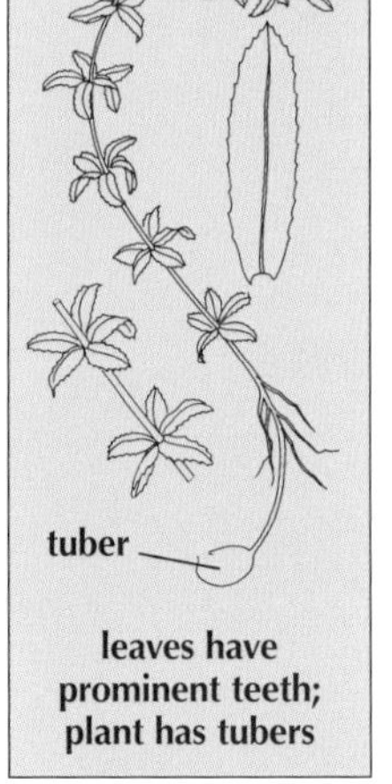

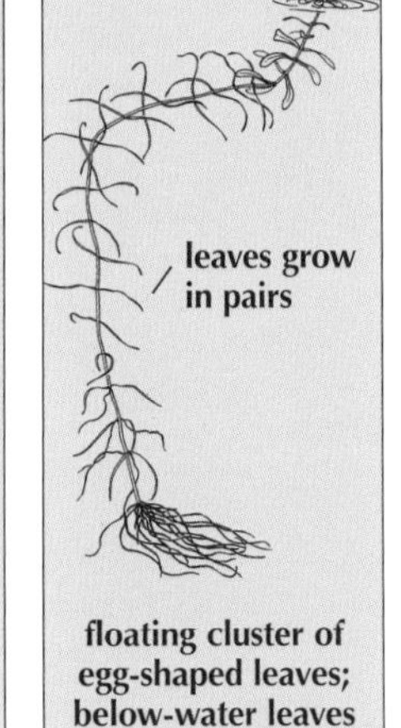

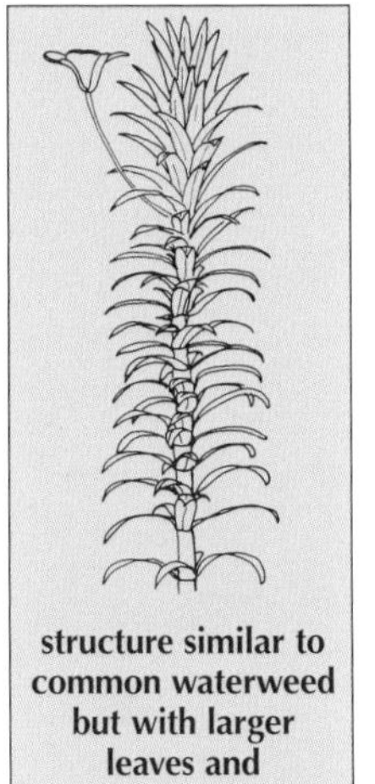

leaves have less prominent teeth; plant has no tubers

leaves have prominent teeth; plant has tubers

floating cluster of egg-shaped leaves; below-water leaves are threadlike

structure similar to common waterweed but with larger leaves and thicker stems

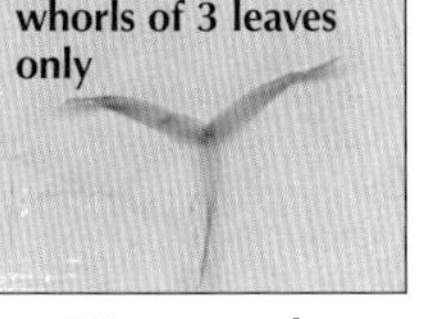

egg-shaped upper leaves

upper leaves in whorls of 4 to 8

Waterweeds
***Elodea* spp., p. 32**

Hydrilla
***Hydrilla verticillata*, p. 30**

Water Starwort
***Callitriche* spp., p. 18**

Brazilian Waterweed
Egeria densa

The two waterweeds, common (*Elodea canadensis*) and Nuttall's (*E. nuttallii*), are hard to distinguish from each other, and can also be mistaken for several similar species, including hydrilla (*Hydrilla verticillata*), water starwort (*Callitriche* spp.), and Brazilian waterweed (*Egeria densa*).

Hydrilla can be distinguished from the two waterweeds by its more prominent teeth on leaf margins, underground tubers, and leaves that grow in whorls of 3 to 5 (leaves of waterweeds usually grow in whorls of 3 only). Water starwort can be distinguished from the two waterweeds and from hydrilla by leaves that grow in pairs. Brazilian waterweed has a structure similar to two *Elodea* waterweeds but its leaves are larger and stems are thicker. Brazilian waterweed is sold in aquarium and aquatic pond stores, sometimes under the name *Anacharis*. This plant is highly invasive in the San Francisco Bay delta and has the potential to invade low salinity Chesapeake tidal waters.

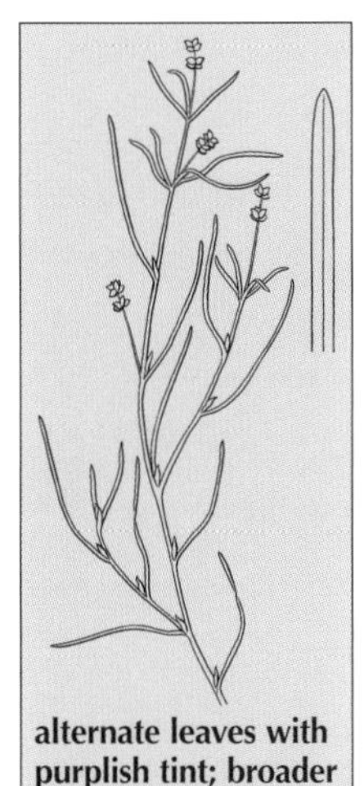

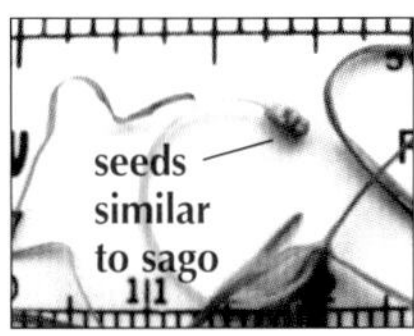

Slender Pondweed, *Potamogeton pusillus,* **p. 20**

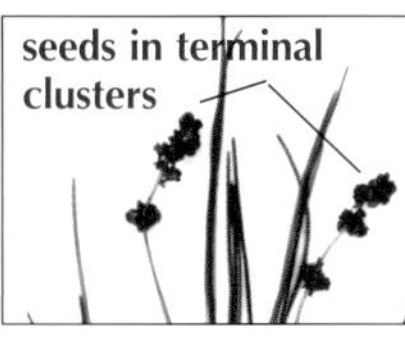

Sago Pondweed *Stuckenia pectinata,* **p. 40**

Horned Pondweed *Zannichellia palustris,* **p. 38**

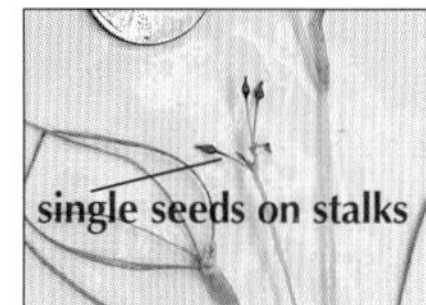

Widgeon Grass *Ruppia maritima,* **p. 42**

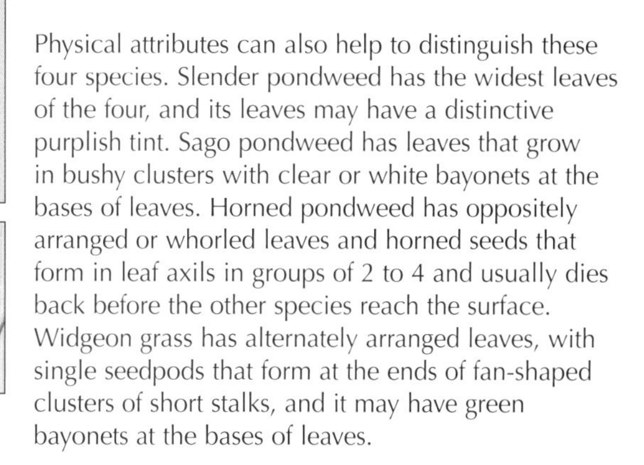

The three species of narrow-leaved pondweeds — slender (*Potamogeton pusillus*), sago (*Stuckenia pectinata*), and horned (*Zannichellia palustris*) — along with widgeon grass (*Ruppia maritima*), can be confused with each other. Luckily for the field guide user, an initial distinction can be made between species based on location. Only horned pondweed and slender pondweed share a fully overlapping distribution area — though in at least one area (the Magothy River) slender pondweed, sago pondweed, and widgeon grass have been found growing together.

Physical attributes can also help to distinguish these four species. Slender pondweed has the widest leaves of the four, and its leaves may have a distinctive purplish tint. Sago pondweed has leaves that grow in bushy clusters with clear or white bayonets at the bases of leaves. Horned pondweed has oppositely arranged or whorled leaves and horned seeds that form in leaf axils in groups of 2 to 4 and usually dies back before the other species reach the surface. Widgeon grass has alternately arranged leaves, with single seedpods that form at the ends of fan-shaped clusters of short stalks, and it may have green bayonets at the bases of leaves.

alternate leaves with rounded tips

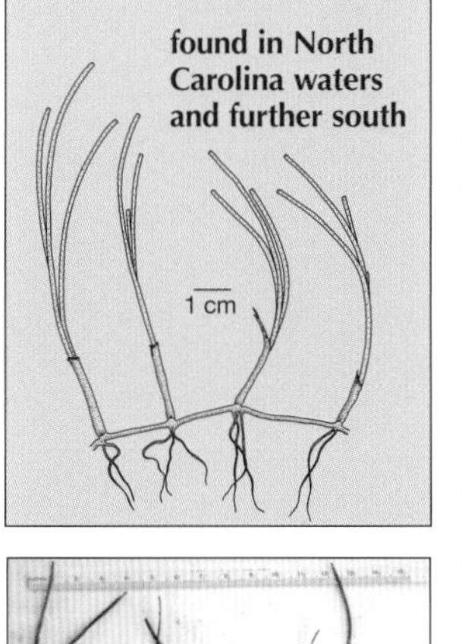

pointed bidentate leaf tip; may have a small central point

found in North Carolina waters and further south

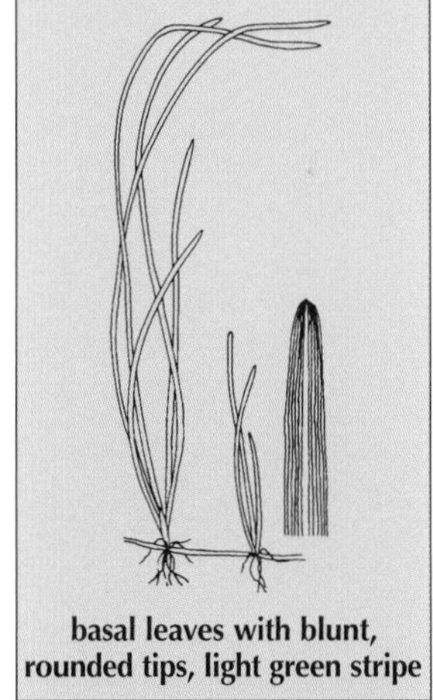

basal leaves with blunt, rounded tips, light green stripe

narrower leaves

Eelgrass
Zostera marina, p. 44

broader leaves; light green stripe

Shoal Grass
Halodule wrightii, p. 46

Wild Celery
Vallisneria americana, p. 24

Eelgrass (*Zostera marina*), shoal grass (*Halodule wrightii*), wild celery (*Vallisneria americana*), and the creeping form of widgeon grass (*Ruppia maritima*) (see p. 51) have similar and sometimes confusing characteristics. Shoal grass is not found in Chesapeake Bay, but may occur in North Carolina waters.

Eelgrass, wild celery, and widgeon grass have rounded leaf tips, while shoal grass has leaves that are pointed, bidentate, and concave. Wild celery has a light green stripe down the middle of its leaves, which are broader than those of the other three. Wild celery also grows in lower salinity than eelgrass or shoal grass, which makes it unlikely that their distributions will overlap.

Photosynthetic life in Chesapeake Bay and Mid-Atlantic coastal waters is rich and diverse. While this guide focuses primarily on SAV, the pages that follow provide information on other plant species you might encounter in these waters. Some of these species can be confused with SAV, while others are quite different but have certain characteristics worthy of interest.

Water Chestnut

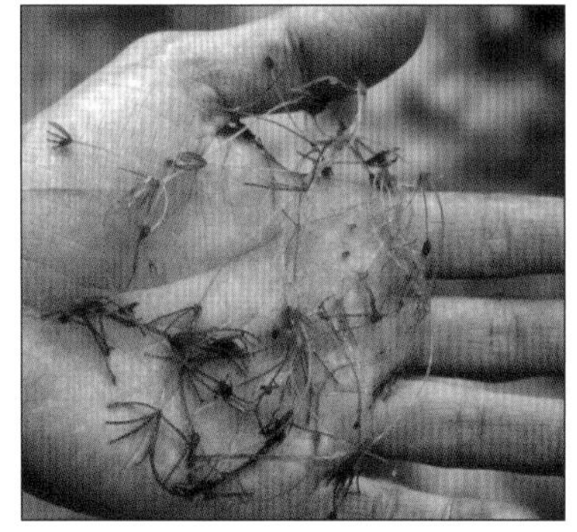

Muskgrass

Cyanobacteria

Water chestnut (*Trapa natans*), for example, belongs to the group of Floating Aquatic Vegetation or FAV. A noxious and invasive plant, it can alter local ecosystems. One acre of water chestnut may produce enough seeds to cover 100 acres the following year, and can out-compete native underwater grasses. The four half-inch spines on the plant's fruit are sharp enough to penetrate shoe leather and large enough to injure waterfowl and keep people off beaches. Water chestnut is included in this guide so you can help quickly identify and report any sightings. For more information see www.dnr.state.md.us/bay/sav/water_chestnut.html.

Several species of algae are abundant in Chesapeake Bay and Mid-Atlantic waters. Muskgrass, for example, can be easily mistaken for SAV. Others, such as Agardh's red weed, are not likely to be confused with SAV, but may be readily encountered.

Blue-green algae (cyanobacteria) are unlikely to be confused with SAV, but they are a fascinating group. Cyanobacteria have provided us with the planet's oldest known fossils — some 3.5 billion years old — and helped to add oxygen to Earth's ancient atmosphere. Not actually plants at all, these photosynthetic bacteria form long, filamentous strands or mat-like structures. They are now often associated with an overabundance of nutrients and low oxygen conditions in coastal waters.

FAV: WATER CHESTNUT

Trapa natans
Family: Hygrocaryaceae

Non-native to Chesapeake Bay; invasive

Recognition: Notorious for its invasiveness and sharp seed pods, water chestnut is classified as floating aquatic vegetation (FAV), not SAV. Its triangular or diamond-shaped leaves form rosettes that float on the water's surface. Leaves, which are 2 to 4 cm long, have a shiny upper side and fine hairs underneath. Leaf margins are toothed, and the petiole or leaf stalks have swollen portions that aid in flotation. Threadlike submersed leaves are arranged alternately on the lower portions of leaf stalks. Water chestnut can be free-floating or firmly rooted to the substrate.

Distribution: Water chestnut is found in both fresh and brackish, slow-moving waters on fine-grained, muddy sediments. It is present throughout the northeastern United States.

Reproduction: Sexual and asexual. Plants flower in June and July, and are pollinated by insects. Inconspicuous white flowers have four sepals, which elongate and harden to form the distinctive barbed spikes of the fruit. While most fruits germinate the following spring, water chestnut seed germination can occur from seeds that are up to 12 years old. Each water

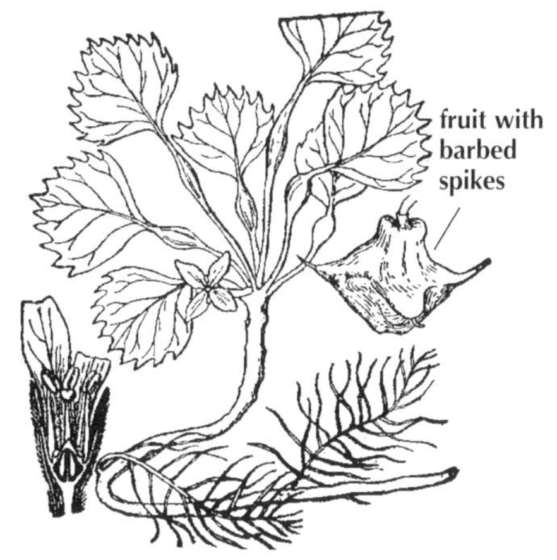

chestnut seed creates 15 to 20 rosettes, and each rosette can generate up to 20 seeds. Asexual reproduction can occur within a growing season if an intact rosette breaks off the parent plant, but rosettes are not able to survive through the winter.

Remarks: Water chestnut was introduced from Eurasia in the 1800s and now occurs in many parts of the eastern United States, clogging waterways and outcompeting native plants. Mechanical and hand harvesting brought water chestnut under control in rivers like the Bird, Bush, and Sassafras in Maryland and today the plant is rare in Chesapeake Bay. Positive identifications of water chestnut should be reported to the Maryland Department of Natural Resources immediately at 410-260-8630.

ALGAE: AGARDH'S RED WEED

Agardhiella spp.
Family: Champiaceae

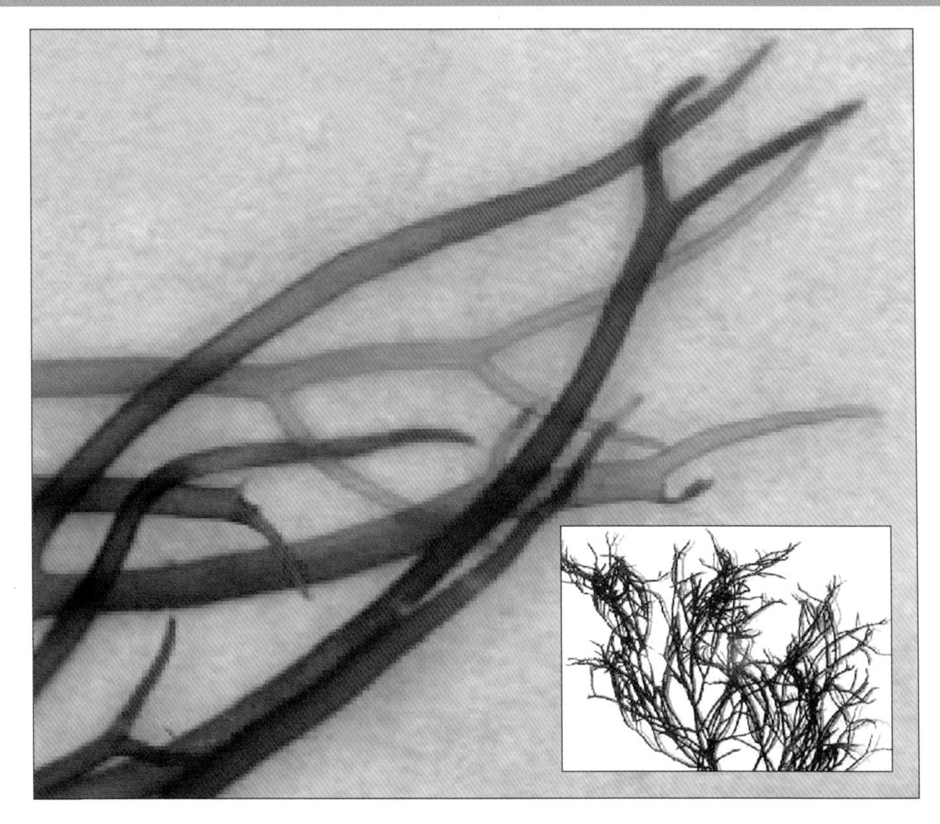

Native to Chesapeake Bay

Recognition: Agardh's red weed is a coarse and bushy alga, and its axis and branches are round. Its branches taper at both the tip and base. Agardh's red weed frequently bleaches to a yellow color in the sun.

Distribution: Agardh's red weed can be found in coastal waters from the tropics north to Cape Cod, including Chesapeake Bay. It occurs most often in shallow, protected, brackish waters (above 15 ppt).

Can be confused with: False agardhiella (*Gracilaria* spp.) and graceful red weed (*Gracilaria* spp.). These two species can be distinguished from Agardh's red weed by their lack of tapering branch bases.

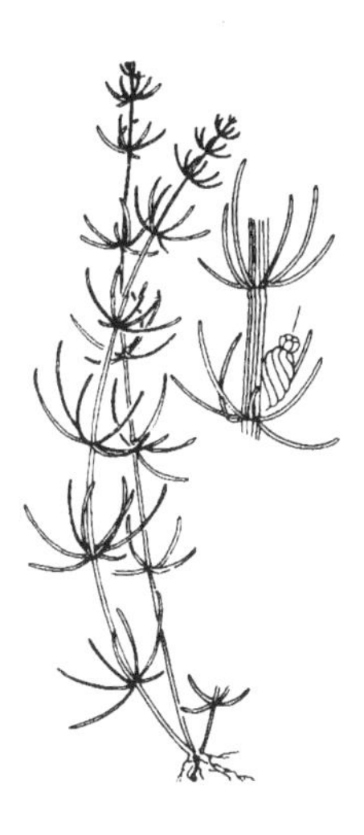

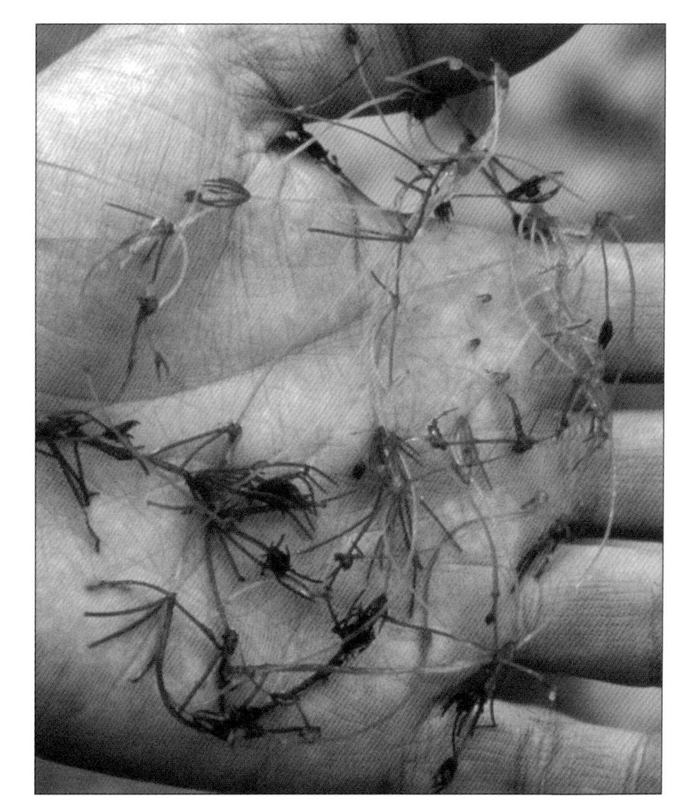

Native to Chesapeake Bay

Recognition: Muskgrass can be mistaken for SAV, but it is a multicellular alga with no true leaves, stems or roots. It is greenish yellow with short, threadlike branches clustering at joints along a stemlike axis. Leaflike structures arise from special nodes in a whorled pattern. Muskgrass anchors to the sediment by branching, rootlike organs called rhizoids. It emits a skunky, musty odor when crushed and feels brittle to the touch.

Distribution: Muskgrass occurs in fresh and brackish water and may be found throughout Chesapeake Bay. It grows in calcium-rich water and on silty or muddy bottoms.

Remarks: The muskgrasses (*Chara* spp.) are an important food for ducks, especially when they bear microscopic, sporelike oogonia.

Can be confused with: Brittle grass (*Nitella* spp.). Another plantlike algae, brittle grass has symmetrically forked branches, unlike muskgrass. Coontail (*Ceratophyllum demersum*) can be distinguished from muskgrass by its lack of odor when crushed or bruised. It is also pliable while muskgrass is brittle.

ALGAE: MUSKGRASS

***Chara* spp. Family: Characeae**
Other names: stoneworts

ALGAE: *ENTEROMORPHA* SPP.

No Common Name
Family: Ulvaceae

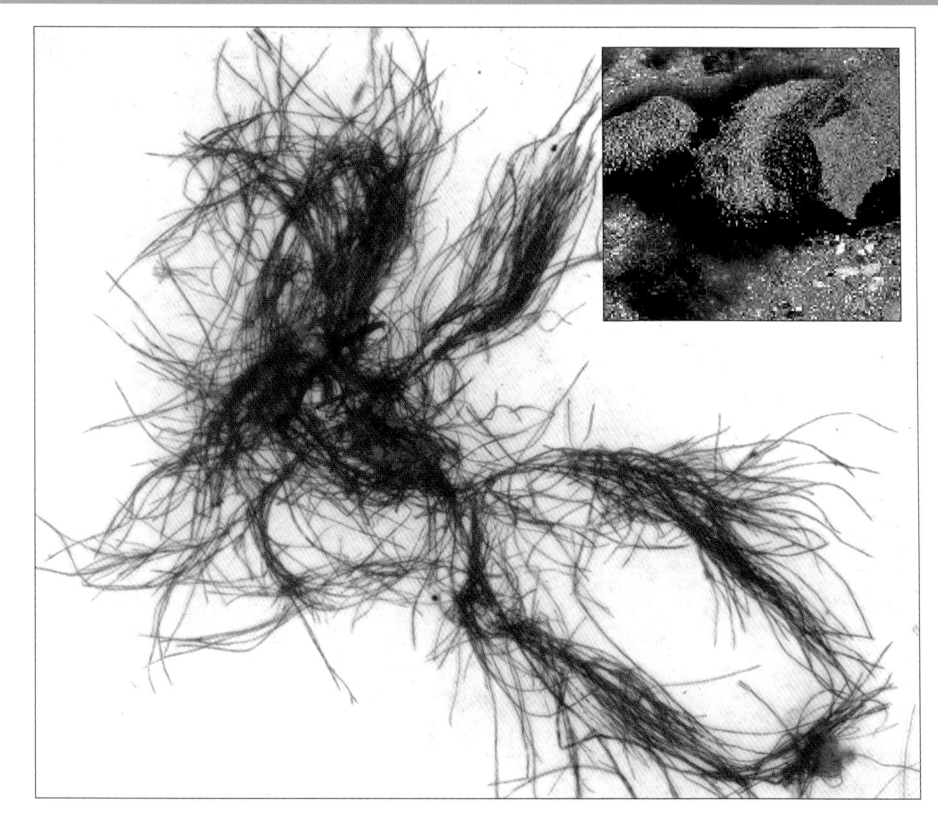

Native to Chesapeake Bay

Recognition: *Enteromorpha intestinalis*, one of the more common species of algae, has thin, tubular, pale green fronds. Air bubbles often appear inside the hollow fronds. Branching may occur. In cross section, *Enteromorpha* species are only two cells thick.

Distribution: At least 11 species of *Enteromorpha* are found throughout shallow waters of Chesapeake Bay. *E. intestinalis* grows abundantly and can form dense meadows in nutrient-rich waters in salinity ranges from 5 ppt to full-strength sea water.

Can be confused with: Eelgrass (*Zostera marina*). Eelgrass can be distinguished from *Enteromorpha* by its darker green color and multiple leaves per shoot. Certain species of *Enteromorpha* may resemble fragments of sea lettuce (*Ulva lactuca*). Some researchers, in fact, have argued that the genus *Enteromorpha* and the genus *Ulva* are one and the same. See for example: www.mbari.org/staff/conn/botany/greens/ram/classification.htm.

Native to Chesapeake Bay

Recognition: False agardhiella is a highly branched red alga with branches that taper at the tips. It is often found floating freely in large clumps over shallow, muddy bottoms.

Distribution: False agardhiella is common south of Cape Cod and is likely to be found in Chesapeake Bay. Another species of algae in this genus, the graceful red weed, can tolerate more brackish water (15 ppt and above) but is more commonly found north of Cape Cod.

Can be confused with: Agardh's red weed (*Agardhiella* spp.). Agardh's red weed can be distinguished from false agardhiella by its branches that taper at both the tip and base.

ALGAE: FALSE AGARDHIELLA

***Gracilaria* spp.**
Family: Solieriaceae

ALGAE: SEA LETTUCE

Ulva lactuca
Family: Ulvaceae

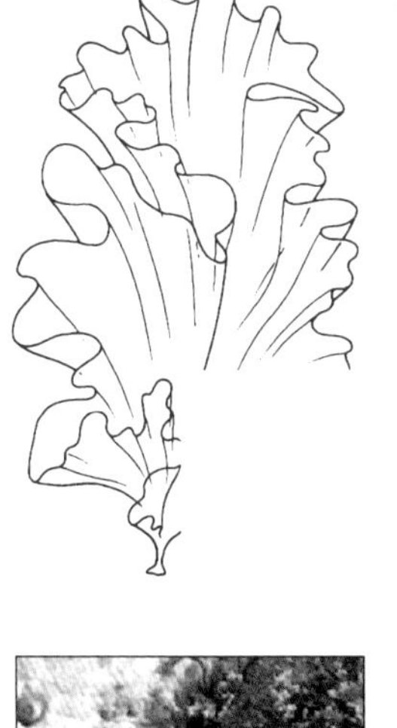

Native to Chesapeake Bay

Recognition: Sea lettuce resembles leaves of wilted green lettuce. Thin and crinkly, it usually grows in a large mass. Leaf cross-sections are one cell thick.

Description: Commonly referred to as seaweed, sea lettuce grows in brackish to high-salinity water throughout Chesapeake Bay (5 ppt to full-strength sea water), especially in nutrient-rich waters. Sea lettuce attaches to rocks, pilings, and other substrates. It can also be free-floating and may wash ashore in large drifts after storms.

Remarks: When abundant, sea lettuce can smother SAV.

Microcystis aeruginosa

Blue-greens (cyanobacteria) and blooms

If you find yourself surrounded by clumps of black and green scum or strangely colored water (brown, red, green, or blue-green), you could be witnessing a bloom of cyanobacteria or phytoplankton. In Chesapeake Bay and its tributaries, blooms tend to occur in the spring and summer. As blooms die off and decompose, they can severely reduce the oxygen available to living resources, and some species can release harmful toxins.

Phytoplankton, such as diatoms or dinoflagellates, are not visible to the naked eye but will discolor the water when they bloom at high density. *Prorocentrum minimum* or *Karlodinium micrum* are two species of dinoflagellate that commonly occur in Chesapeake Bay. For more information on harmful algal blooms (HABs) see www.dnr.state.md.us/bay/hab/ or www.whoi.edu/redtide/.

Many species of cyanobacteria, in contrast, grow in colonies or chains that can resemble species of macroalgae. This is why cyanobacteria are also called blue-green algae, even though they are bacteria — not algae at all. Below are several species of cyanobacteria that you might find in Chesapeake Bay. For

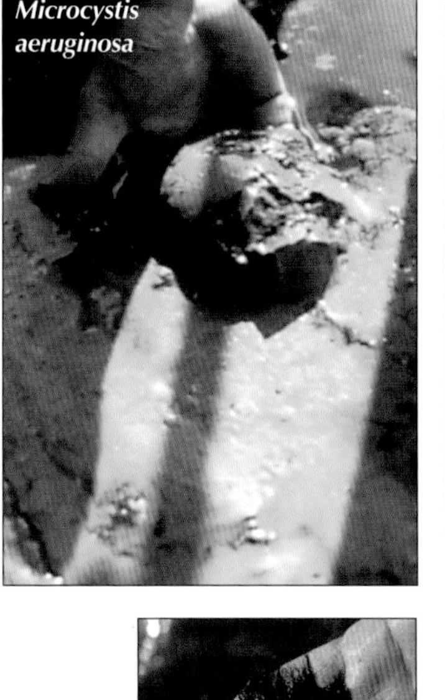

Microcystis aeruginosa

Oscillatoria

Microcystis aeruginosa

more information see: www.dnr.state.md.us/bay/cblife/algae/cyano/index.html

***Anabaena* spp.:** These species can be found in fresh water, predominantly in the Potomac (above Indian Head) and in the upper Bay. These are filamentous cyanobacteria, capable of nitrogen fixation. *Anabaena* can produce toxins that may cause skin irritations or nausea and can impart noxious odors and disagreeable tastes to the water.

Microcystis aeruginosa: Found in fresh water, predominantly in the Potomac (above Indian Head) and in the upper Bay. Comprised of small cells embedded in a gelatinous matrix, cells 3 to 4.5 micrometers in diameter. During heavy blooms, the water may appear like blue-green paint billowing near the surface. Some strains of *Microcystis* may produce toxins reported to cause health problems for animals that drink the water, as well as minor skin irritation and gastrointestinal discomfort in humans.

Oscillatoria: Occurs in moderate salinity to fresh water in the mesohaline area of the Bay and the Potomac. A very small filament (0.6 to 1.0 micrometers), in streams this algae looks almost black in appearance.

Why is SAV Important?

- Underwater grasses (submerged aquatic vegetation or SAV) shelter spawning fish and shellfish and their offspring, such as juvenile blue crabs.
- Their seeds and tubers provide the primary food source for many waterfowl that over-winter on the Chesapeake and coastal bays.
- Clams, oysters, and other filter-feeding organisms use decaying bits of underwater grasses for food.
- Underwater grasses help keep the water clean by trapping sediment and absorbing nutrients that can lead to algae overgrowth, cloudy water, and low dissolved oxygen — including so-called "dead zones" where fish and plants cannot live.
- Plants produce oxygen and convert carbon, nitrogen, and phosphorus into organic matter that living things can use for energy and growth.

- Beds of SAV slow water currents and reduce the force of waves, which protects against shoreline erosion.
- SAV helps to anchor the substrate so other organisms can grow and colonize.

SAV Habitat Requirements

Because SAV requires light for photosynthesis, it grows in shallow areas where sufficient light can penetrate the water. In the Chesapeake Bay region, this normally means water that is less than 6 feet (2 meters) deep at low tide. Certain species of SAV can tolerate lower light levels, allowing them to grow in deeper water or in cloudy or muddy conditions.

Scientific studies have identified specific components in the water column that reduce the amount of light reaching the plants. These include total suspended solids, algae (measured as chlorophyll *a*), dissolved inorganic nitrogen, and dissolved inorganic phosphorus. Threshold levels for these components and other water quality parameters that must be met for SAV to grow have been set by the Environmental Protection Agency's Chesapeake Bay Program. These parameters, called "SAV habitat requirements," can be found on the Chesapeake Bay Program web site at: www.chesapeakebay.net/pubs/sav/index.html.

Charting SAV in the Bay

Submerged Aquatic Vegetation in Chesapeake Bay

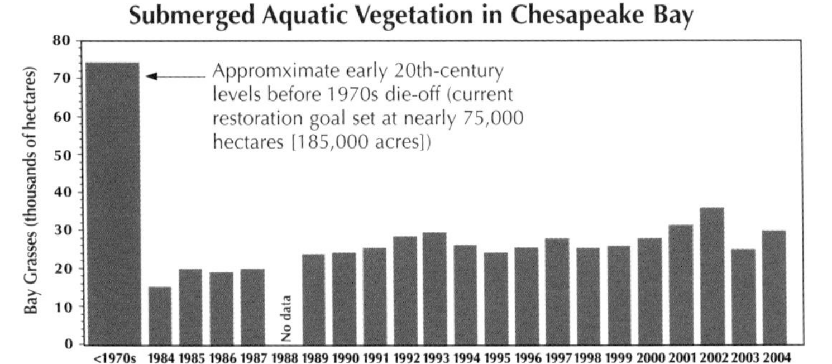

Source: Adapted from the Chesapeake Bay Program and Moore et al. (2004).

Aerial photographic surveys of Chesapeake Bay and nearby coastal waters have measured the abundance and distribution of SAV in the region since the early 1970s. These surveys, conducted by the Virginia Institute of Marine Science (VIMS) along with various partners, have helped to develop data that can be used to monitor year-to-year changes and to gauge the success of SAV restoration efforts in the Bay.

From May to November each year, these aerial surveys photograph about 240,000 hectares of Chesapeake Bay. The flight lines are positioned to include all areas known to have SAV, as well as most areas less than two meters deep in the middle and upper zones of the Bay, where SAV could grow. From 1995-2004, SAV coverage has averaged about 28,000 hectares.

The chart above summarizes the results of the annual survey and shows the total hectares of SAV in Chesapeake Bay from 1984-2004. The *Chesapeake 2000* agreement set a restoration goal of nearly 75,000 hectares (185,000 acres) of SAV coverage by the year 2010. Even at a two-decade high in 2002 (36,283 hectares), SAV coverage accounted for only 48 percent of the 2010 restoration goal.

You can follow the trends in SAV recovery in Chesapeake Bay and find updated information each year at www.vims.edu/bio/sav/index.html.

Saving SAV

SAV populations in Chesapeake Bay have been declining for the last 200 years, but the most widespread and drastic decline occurred in the late 1960s and early 1970s. That decline continued through the mid-1980s, and then SAV coverage slowly began to increase. The increase remained fairly steady through 1993, but has fluctuated since, generally going up in years with low precipitation due to better water quality, and down in high-flow years due to worse water quality. For example, 2003, a high-flow year, saw a 30 percent drop in SAV from the relative high seen in 2002, a year of low flow. (See: www.chesapeakebay.net/status.cfm?sid =88.)

The general decline of SAV is linked to degraded water quality caused by development, storms, runoff, and pollution from excess nutrients entering the Bay from sewage plants, farms, fossil fuel combustion, and fertilized lawns. Although scientists and resource managers generally agree that reduced light penetration associated with human population growth and development is the main cause of widespread SAV loss, the remaining populations are also at risk from hydraulic dredges used to harvest clams, scarring and damage from boat propellers, and other human disruptions. A variety of efforts are underway to conserve and restore SAV, including the following:

- Improving water quality by reducing nutrient pollution, especially the amount of nitrogen and phosphorus reaching the Bay.
- Reducing manmade physical disturbances in shallow, nearshore waters where SAV

grows, through improved education and regulations, such as teaching boaters how to avoid harming plants with propellers and prohibiting clam harvesting in SAV beds.

- Reseeding SAV to supplement natural recolonization, an effort that has already restored limited amounts of SAV habitat. Most important, however, is maintaining existing SAV populations, since this is the key to increasing the area of SAV in the Bay.

A good source of additional information on techniques for conserving and restoring SAV is *Guidelines for Conservation and Restoration of Seagrass in the United States and Adjacent Waters* by Fonseca and others. For details on this book and other useful information, see References (p. 73).

What You Can Do to Conserve SAV

- Reduce the amount of fertilizers applied to yards.
- Plant native vegetation suited to your soil, moisture, and climate conditions.
- If you need to fertilize, follow all directions carefully; never apply before storms.
- Reduce shoreline erosion by planting appropriate shoreline vegetation.
- If you have access to shallow-water areas, volunteer to survey SAV in your area.
- When boating, avoid disturbing SAV beds. Propellers may tear rooted vegetation out of bottom sediments.
- If you own waterfront property, avoid using herbicides that may harm delicate SAV plants.

The U.S. Fish and Wildlife Service supplied the tips above. For more information see the Chesapeake Bay Field Office website at www.fws.gov/chesapeakebay/CBSAV/HTM.

To help pinpoint the SAV species you find in and around Chesapeake Bay, these maps chart *potential* habitat by salinity range. Maps show salinity ranges in red where each species grows best (see charts on pp. 6-7 for full growth range). Salinity can vary from year to year, and plant growth also depends on other factors. For up-to-date maps of *actual* grass distribution, visit the web at www.vims.edu/bio/sav/groundsurveymaps.html.

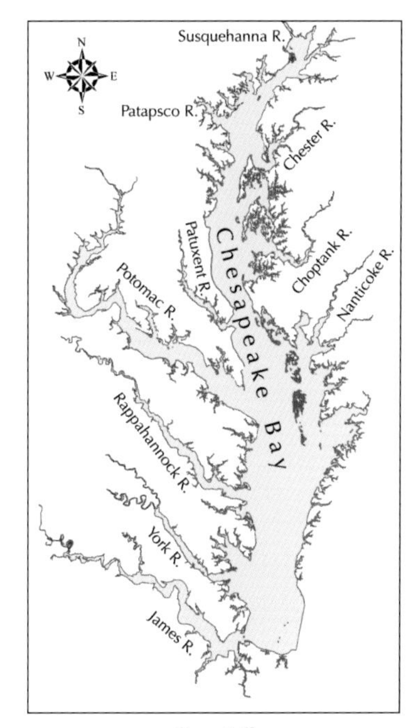

Coontail
(*Ceratophyllum demersum*)
0 to 6.5 ppt
low to medium salinity

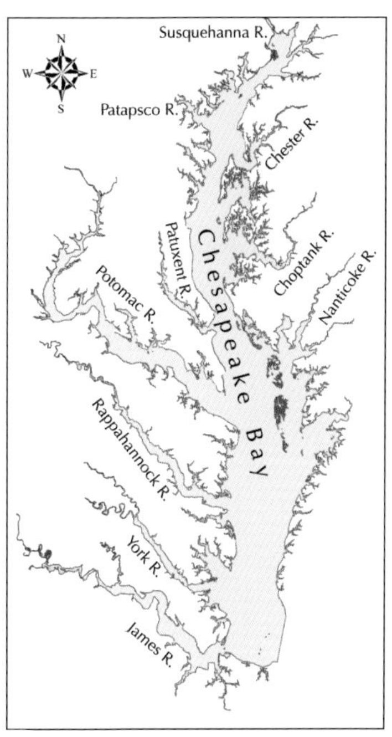

Water Starwort
(*Callitriche* spp.)
0 to 2 ppt
low salinity

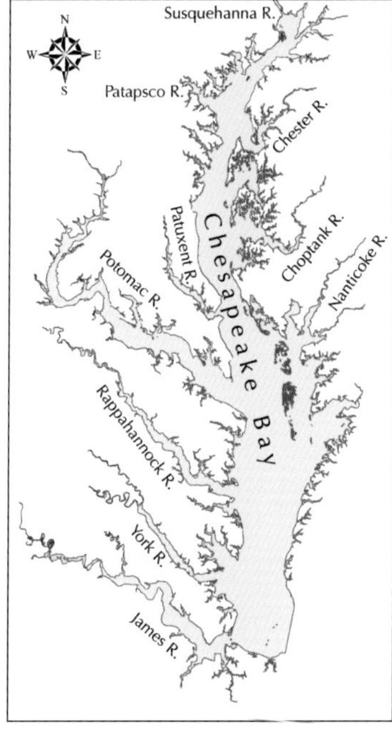

Slender Pondweed
(*Potamogeton pusillus*)
0 to 4 ppt
low salinity

POTENTIAL SAV HABITAT IN CHESAPEAKE BAY

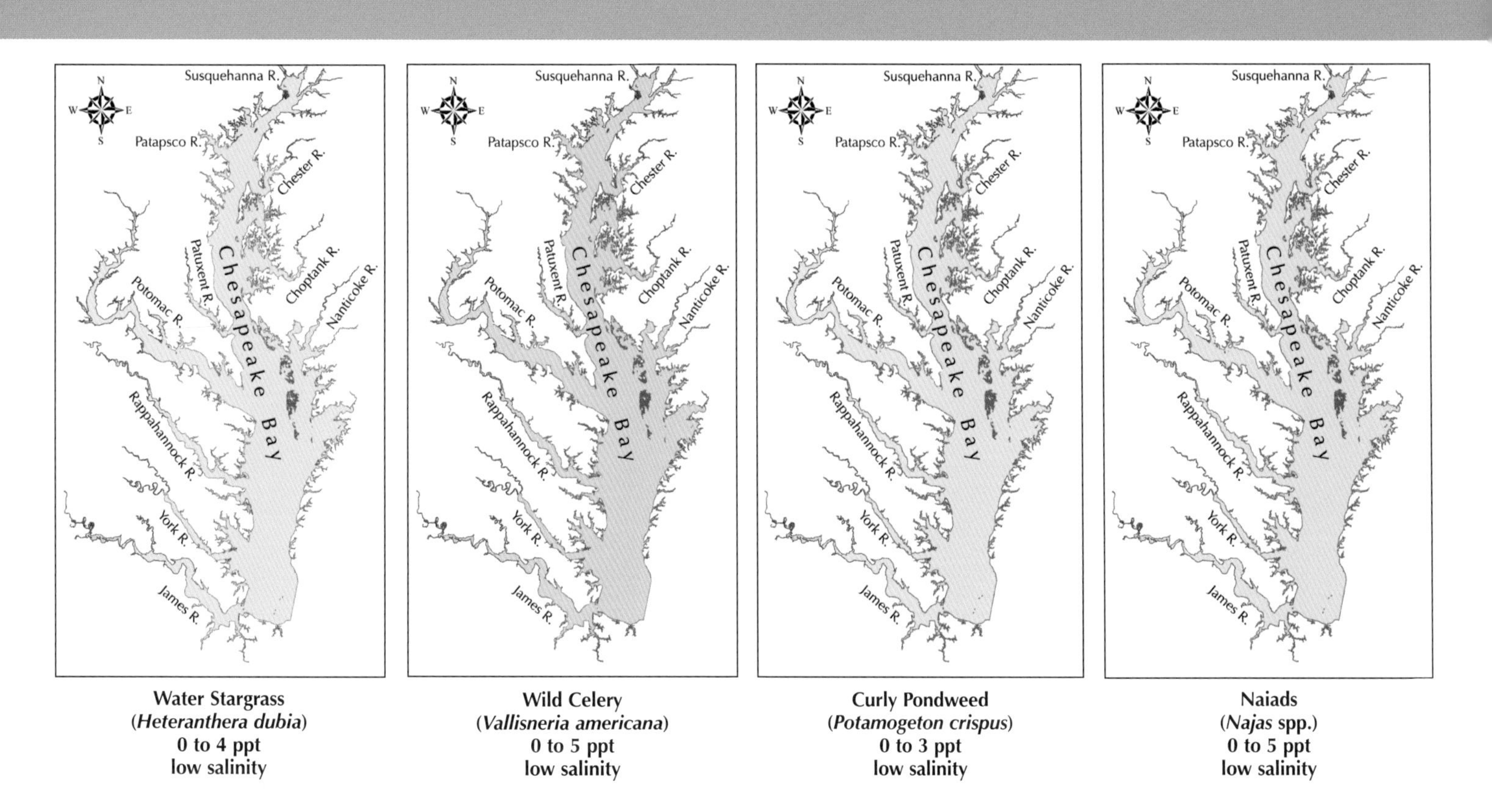

Water Stargrass
(*Heteranthera dubia*)
0 to 4 ppt
low salinity

Wild Celery
(*Vallisneria americana*)
0 to 5 ppt
low salinity

Curly Pondweed
(*Potamogeton crispus*)
0 to 3 ppt
low salinity

Naiads
(*Najas* spp.)
0 to 5 ppt
low salinity

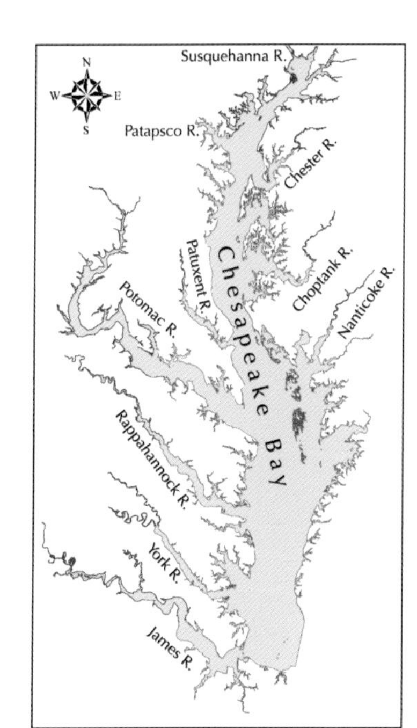

Hydrilla
(*Hydrilla verticillata*)
0 to 5 ppt
low salinity

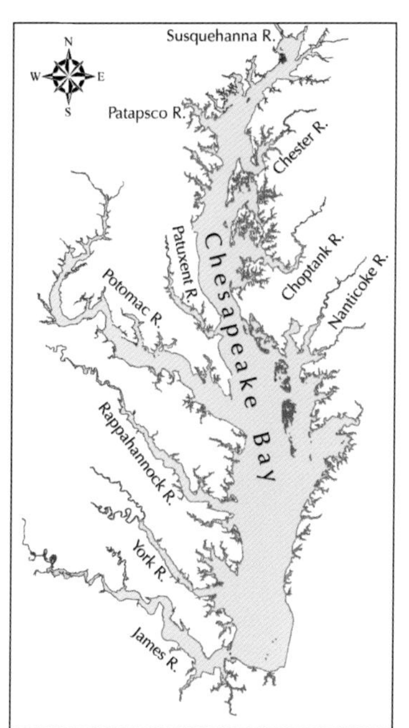

Waterweeds
(*Elodea* spp.)
0 to 4 ppt
low salinity

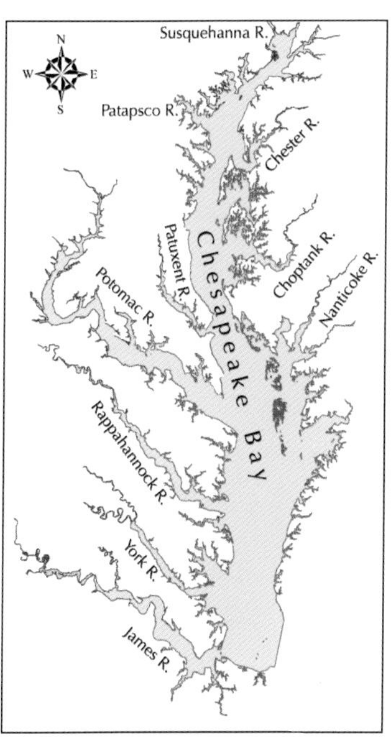

Eurasian Watermilfoil
(*Myriophyllum spicatum*)
0 to 5 ppt
low salinity

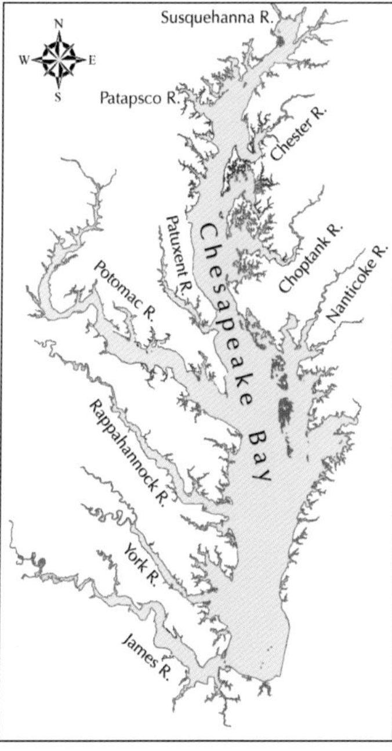

Redhead Grass
(*Potamogeton perfoliatus*)
5 to 10 ppt
medium salinity

POTENTIAL SAV HABITAT IN CHESAPEAKE BAY

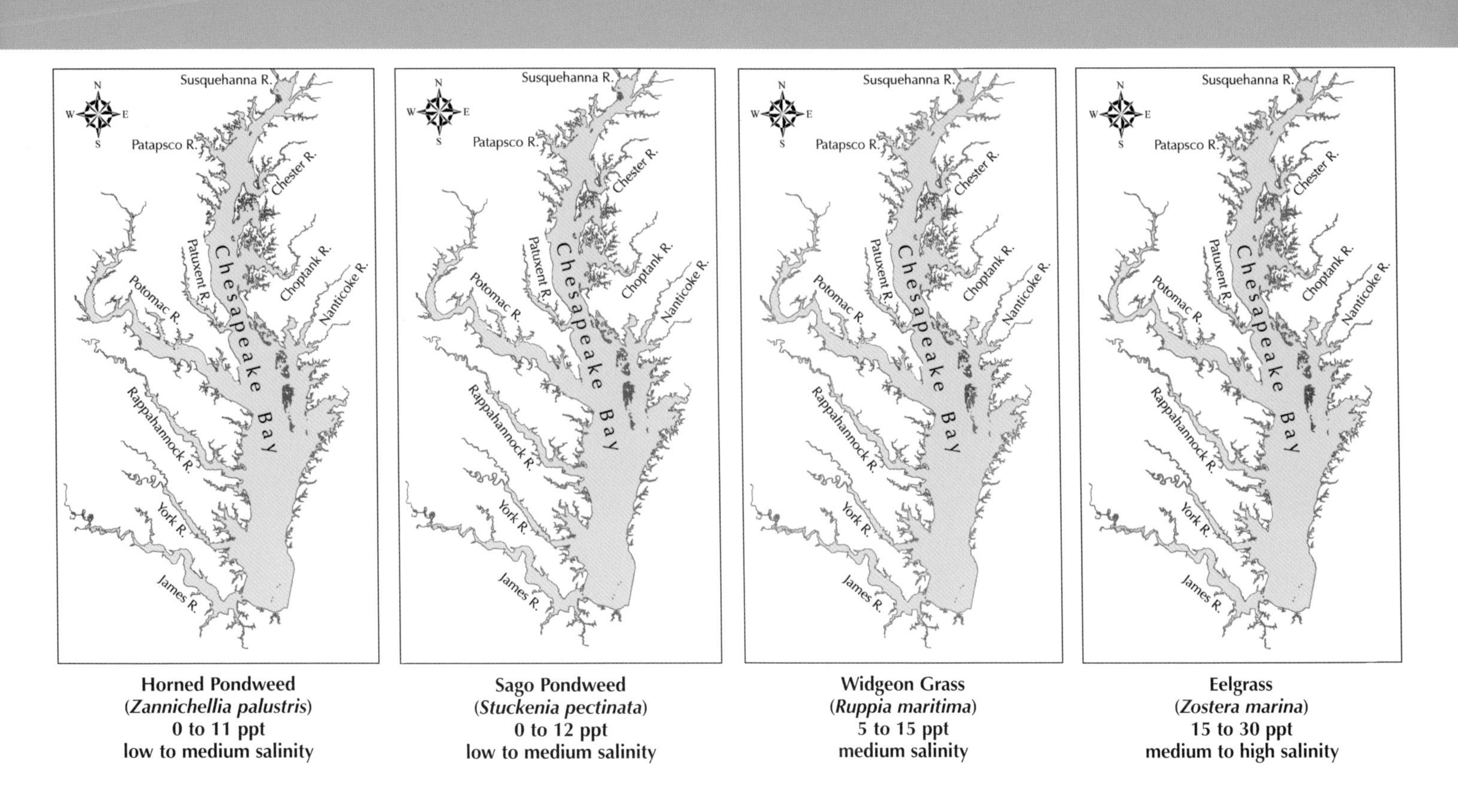

Horned Pondweed
(*Zannichellia palustris*)
0 to 11 ppt
low to medium salinity

Sago Pondweed
(*Stuckenia pectinata*)
0 to 12 ppt
low to medium salinity

Widgeon Grass
(*Ruppia maritima*)
5 to 15 ppt
medium salinity

Eelgrass
(*Zostera marina*)
15 to 30 ppt
medium to high salinity

asexual reproduction — expansion of the leaves, stems, roots, and rhizomes of the plant by cell division

adventitious — developing in an irregular or uncharacteristic position, especially in reference to buds and roots

alternate — leaves not opposite to each other but arranged singly along a stem

annual — a plant that completes its life cycle in one year

antherida — male sexual part in algae

axil — angle usually formed by a leaf or petiole with the stem from which it arises

axillary — in or associated with the axil

basal — at the base or bottom of a plant

bisexual — a perfect flower (both stamens and pistil)

bract — modified leaf associated with, but not part of, a flower

bulbils — vegetative (asexual) bud produced at stem nodes in algae

calcareous — with a high concentration of calcium carbonate

cuticle — protective, waterproof, waxy layer covering plant surfaces

dioecious — male and female flowers occurring on separate plants

entire — without visible teeth or divisions (refers to leaf margins)

eutrophic — containing a high concentration of nutrients; particularly nitrogen and phosphorus

fibrous — composed of or resembling fibers

filiform — threadlike

frond — leaflike thallus

hydroid — highly branched, colonial group of invertebrate organisms, closely related to sea anemones

hypanthium — an enlarged or developed flower receptacle

imperfect flowers — having pistils and stamens on separate flowers

inflorescence — the entire flower cluster

invertebrate — animal without a backbone, such as insects and worms

lanceolate — shaped like a lance or arrowhead

lateral — on the sides

linear — long and narrow with parallel sides

meristem — embryonic plant tissue that is actively dividing, such as is found at the tip of stems and roots

monoecious — male and female flowers occurring on the same plant

node — positions on upper stems, usually bearing leaves, or on lower stems, usually bearing roots

obovate — egg-shaped, with the broader end above the middle (refers to leaves)

oogonia — the sex organ of some algae and fungi that contains female gametes (oospores)

opposite — leaves arranged directly across from one another along a stem

ovoid — solid, with an egg-shape

peduncle — flower stalk

perennial — a plant living more than two years

perfect flower — having both stamens and pistil (bisexual)

perianth — petals and sepals together

petals — inner leaves of a flower

petiole — the lower stalk of a leaf

pinnate — compound leaves arranged on both sides of a common axis, as in a feather

pistil — seed-bearing organ of a female flower

pistillate — containing pistils

pollen — reproductive grains contained in stamens (male)

ppt — parts per thousand (the units of measure for salinity)

rachis — main axis of a leaf or spike

rhizoids — slender, root-like organs

rhizome — lower horizontal stems either prostrate on sediment surface or buried; usually with roots and new shoots at stem nodes and curving upward at the ends

rosette — circular cluster of leaves or other structures

runners — branches off buried rhizomes; usually with tubers produced at the end

sepals — the outer leaves of a flower

serrated — having visible sharp teeth (refers to leaf margins)

sheath — enveloping lower parts of leaves

spathe — a large bract enclosing a flower or group of flowers

spike — a large group of flowers on an elongated axis

stamen — male organ of a flower, composed of a slender stalk (filament) with a pollen-bearing anther at the end

staminate — containing stamens

stipule — sheath or appendage (usually in pairs) at the base of a leaf or its petiole

stolon — prostrate (lying flat on the ground), slender, aboveground stem producing new plants at nodes

stomata — openings on the epidermis of a leaf or stem through which water vapor moves out of the plant and carbon dioxide moves in

tendril — a slender, clasping, or twining outgrowth

terminal — at the tips or end

truncate — with the base or tip transversely straight as if cut off

tubers — vegetative (asexual) buds (buried in the sediment) usually forming at the ends of runners; capable of remaining dormant before developing into new plants

turions — vegetative (asexual) buds usually formed in the leaf axils or stem tips; capable of remaining dormant before developing into new plants

undulated — having a wavy margin or surface

vegetative reproduction — see asexual reproduction

whorl — a circle of 3 or more branches, leaves or flower stalks arising from the same node

Batiuk, R.A., R.J. Orth, K.A. Moore, W.C, Dennison, J.C. Stevenson, L.W. Staver, V. Carter, N.B. Rybicki, R,E. Hickman, S. Kollar, S. Bieber, and P. Heasly. 1992. Submerged Aquatic Vegetation Habitat Requirements and Restoration Target: A Technical Synthesis. U.S. Environmental Protection Agency, Chesapeake Bay Program, CBP/TRS 83/92, Annapolis, MD.

Batiuk, R.A., P. Bergstrom, M. Kemp, E. Koch, L. Murray, J.C. Stevenson, R. Bartleson, V. Carter, N.B. Rybicki, J.M. Landwehr, C. Gallegos, L. Karrh, M. Naylor, D. Wilcox, K.A. Moore, S. Ailstock, and M. Teichberg. 2000. Chesapeake Bay Submerged Aquatic Vegetation Water Quality and Habitat Based Requirements and Restoration Targets: A Second Technical Synthesis. Chesapeake Bay Program, Annapolis, MD. Available online at: www.chesapeakebay.net/pubs/ sav/index.html.

Beal, E.O. 1977. A manual of marsh and aquatic vascular plants of North Carolina with habitat data. North Carolina Agricultural Research Service, Raleigh, NC.

Brown, M.L. and R.G. Brown. 1984. Herbaceous Plants of Maryland. Port City Press, Baltimore, MD. 1127 pp.

Carter, V., P.T. Gammon, and N.C. Bartow. 1983. Submersed aquatic plants of the tidal Potomac River; U.S. Geological Survey Bulletin 1543.

Dennison, W.C., R.J. Orth, K.A. Moore, J.C. Stevenson, V. Carter, S. Kollar, P.W. Bergstrom, and R.A. Batiuk. 1993. Assessing water quality with submersed aquatic vegetation. Bioscience 43(2):8694.

Fassett, N.C. 1957. A Manual of Aquatic Plants. University of Wisconsin Press. Madison, WI. 405 pp.

Fernald, M.L. 1970. Gray's Manual of Botany. D. Van Nostrand Company, New York, NY. 1632 pp.

Fonseca, M.S., W.J. Kenworthy, and G.W. Thayer. 1998. Guidelines for the Conservation and Restoration of Seagrasses in the United States and Adjacent Waters. NOAA Coastal Ocean Program, Decision Analysis Series No. 12. Silver Spring, MD. 222 pp.

Haynes, R.R. 2000. Hydrocharitaceae. In Volume 22, Flora of North America, accessed from www.efloras.org/florataxon.aspx?flora_id=1&taxon_id=10426. Publishing information: www.oup.com/us/catalog/general/subject/LifeSciences/Botany/?view=usa&ci=0195137299.

Hotchkiss, N. 1967. Underwater and Floating-leaved Plants of the United States and Canada. Patuxent Wildlife Research Center, Dover Publications, Inc., New York, NY, Resource Publication 44. 124 pp.

Humm, H.J. 1979. The Marine Algae of Virginia. Published for the Virginia Institute of Marine Science by University Press of Virginia. Charlottesville, VA. 263 pp.

Hurley, L.M. 1990. Field Guide to the Submerged Aquatic Vegetation of Chesapeake Bay. U.S. Fish and Wildlife Service. Annapolis, MD. 51 pp.

Korscbgen, C.E. and W.L. Green. 1988. American Wild Celery (*Vallisneria americana*): Ecological Consideration for Restoration. U.S. Fish and Wildlife Service, Fish and Wildlife Technical Report 19. 24 pp.

Moore, K.A., D. J. Wilcox, and R.J. Orth. 2000. Analysis of the abundance of submerged aquatic vegetation communities in the Chesapeake Bay. Estuaries 23(l):115-127.

Moore, K.A., R.L. Wetzel, and R.J. Orth. 1997. Seasonal pulses of turbidity and their relations to eelgrass (*Zostera marina*) survival in an estuary. J. Exp. Mar. Bio. Ecol. 215:115-134.

Smithsonian Marine Station at Fort Pierce, FL. www.sms.si.edu/IRLFieldGuide/Halodu_beaudet.htm.

Stevenson, J.C. and N.M. Confer, eds. 1978. Summary of available information on Chesapeake Bay submerged vegetation. U.S. Fish and Wildlife Service, Office of Biological Services. FWS/OBS-78/66. 335 pp.

Stevenson, J.C., L.W. Staver, and K.W. Staver. 1993. Water quality associated with survival of submersed aquatic vegetation along an estuarine gradient. Estuaries 1(2):346-361.

Tiner, R.W., Jr. 1987. A Field Guide to Coastal Wetland Plants of the Northeastern United States. University of Massachusetts Press. Amherst, MA. 285 pp.

United States Environmental Protection Agency. 1995. Guidance for Protecting Submerged Aquatic Vegetation in Chesapeake Bay from Physical Disruption. Chesapeake Bay Program CBP/TRS 139/95. Annapolis, MD.

White, C.P. 1989. Chesapeake Bay: Nature of the Estuary, A Field Guide. Tidewater Publishers, Centreville, MD. 211 pp.

PHOTO AND ILLUSTRATION CREDITS

Note: Photographer's/illustrator's affliation and or publication source listed in full with first entry only.

Cover: Large photograph by Erica Goldman, Maryland Sea Grant; small photograph by Linda M. Hurley, U.S. Fish & Wildlife Service/Department of Interior (USFWS/DOI).

Inside: U.S. Map, mapresources.com.

p. 2 All photos, Michael D. Naylor, Maryland Department of Natural Resources (MDNR).

p. 3 Top left, Michael D. Naylor; bottom left, Wayne Carmichael (Wright State University), Mark Schneegurt (Wichita State University), and Cyanosite (www.cyanosite.bio.purdue.edu).

p. 4 Map, Paula Jasinski and Tara Shleser, National Oceanic and Atmospheric Administration (NOAA) Chesapeake Bay Office, Gloucester Point, VA.

p. 5 Maps, David Jasinski, University of Maryland Center for Environmental Science.

pp. 8-15 SAV Identification key designed by Michael D. Naylor, MDNR; drawings by Sedgewick Cole, MDNR.

p. 9 Left, Peter W. Bergstrom, NOAA/Department of Commerce (NOAA/DOC); middle, Ronald C. Phillips, University of Hawai'i at Manoa (UHM); right, Linda M. Hurley, USFWS/DOI.

p. 10 Left, Linda M. Hurley, USFWS/DOI; right, MDNR.

p. 11 All photos, MDNR.

p. 12 Left, MDNR; right, Peter W. Bergstrom, NOAA/DOC.

p. 13 Left and middle, Linda M. Hurley, USFWS/DOI; right, MDNR.

p. 14 Left and middle, Peter W. Bergstrom, NOAA/DOC; right, Steven Ailstock, Anne Arundel Community College (AACC).

p. 15 Left and left middle, MDNR; right middle and right, Linda M. Hurley, USFWS/DOI.

p. 16 Peter W. Bergstrom, NOAA/DOC.

p. 17 Drawing, Karen Teramura, USFWS/DOI; top, MDNR; bottom left, Linda M. Hurley, USFWS/DOI; bottom right, MDNR.

p. 18 MDNR.

p. 19 Drawing by Rebecca A. Haefner, used with permission (in Whitley, James R., Barbara Basset, Joe G. Dillard and Rebecca A. Haefner. 1990. Water Plants for Missouri Ponds, Missouri Department of Conservation, Jefferson City, MO); top left and right, MDNR; bottom left, Peter W. Bergstrom, NOAA/DOC.

p. 20 Peter W. Bergstrom, NOAA/DOC.

p. 21 Drawing, John Norton; top and bottom left, Peter W. Bergstrom, NOAA/DOC; bottom right, MDNR.

p. 22 Peter W. Bergstrom, NOAA/DOC.

p. 23 Drawing, Karen Teramura, USFSWS/DOI; top left, Linda M. Hurley, USFWS/DOI; top right, Glenn Fawcett, © 2004, reprinted with permission from The Baltimore Sun; bottom, Nancy Rybicki, U.S. Geological Survey (USGS).

p. 24 Top, Michael D. Naylor, MDNR; bottom, Mike Haramis, U.S. Geological Survey/Biological Resources Discipline (USGS/BRD).

p. 25 Drawing, Karen Teramura, USFWS/DOI; top left photo, Mike Haramis, USGS/BRD; top right, Michael D. Naylor, MDNR; middle, Steven Ailstock, AACC; bottom left and far right, Peter W. Bergstrom, NOAA/ DOC; bottom second and third from left, Linda M. Hurley, USFWS/DOI; bottom second from right, Mike Haramis, USGS/BRD.

p. 26 Peter W. Bergstrom, NOAA/DOC.

p. 27 Drawing, John Norton; top left, right, and middle, Linda M. Hurley, USFWS/DOI; bottom left, Peter W. Bergstrom, NOAA/DOC.

p. 28 Linda M. Hurley, USFWS/DOI.

p. 29 Drawing of whole plant, Karen Teramura; closeup drawings of leaves used with permission (from Walter Conrad Muenscher. 1944. Aquatic Plants of the United States, Vol. IV, Comstock Publishing Company, Inc., Cornell University. Ithaca, NY); top left, right, and middle, Linda M. Hurley, USFWS/DOI; bottom left, MDNR.

p. 30 Michael D. Naylor, MDNR.

p. 31 Drawing, Karen Teramura; top, MDNR; bottom left, Richard Hammerschlag, USGS Patuxent Wildlife Research Center; bottom middle, Linda M. Hurley, USFWS; bottom right, Michael D. Naylor, MDNR.

p. 32 Linda M. Hurley, USFWS.

p. 33 Drawing, Karen Teramura, USFWS/DOI; top, MDNR; bottom left and right, MDNR.

p. 34 Linda M. Hurley, USFWS/DOI.

p. 35 Drawing, Karen Teramura; top left, Peter W. Bergstrom, NOAA/DOC; top right, MDNR; bottom left and right, Linda M. Hurley, USFWS/DOI.

p. 36 Linda M. Hurley, USFWS/DOI.

p. 37 Drawing, Karen Teramura, USFWS/DOI; top left, MDNR; right, Steven Ailstock, AACC; bottom left, Linda M. Hurley, USFWS/DOI.

p. 38 Michael D. Naylor, MDNR.

p. 39 Drawing, Karen Teramura, USFWS/DOI; left, MDNR; right, Linda M. Hurley, USFWS/DOI.

p. 40 Peter W. Bergstrom, NOAA/DOC.

p. 41 Drawing, Karen Teramura, USFWS/DOI; top left, Steven Ailstock, AACC; right and bottom middle, Linda M. Hurley, USFWS/DOI; bottom left, Mike Haramis, USGS/BRD.

p. 42 Peter W. Bergstrom, NOAA/DOC.

p. 43 Drawing, Karen Teramura, USFWS/DOI; top and bottom left, Peter W. Bergstrom, NOAA/DOC; top right, MDNR; middle, Linda M. Hurley, USFWS/DOI; bottom right, Mike Haramis, USGS/BRD.

p. 44 Peter W. Bergstrom, NOAA/DOC.

p. 45 Drawing, Karen Teramura, USFWS/DOI; top and bottom left, Ronald C. Phillips, UHM; top and bottom right, Robert F. Murphy, Alliance for the Chesapeake Bay (ACB).

p. 46 Ronald C. Phillips, UHM.

p. 47 Drawing and photos, Ronald C. Phillips, UHM (drawing only in Ronald C. Phillips. 1988. Seagrasses. Smithsonian Contributions to the Marine Sciences, Number 34. Smithsonian Institution Press, Washington, DC).

p. 48 Left, Peter W. Bergstrom, NOAA/DOC; right, Linda M. Hurley, USFWS/DOI.

p. 49 Two left drawings, Karen Teramura, USFWS/DOI; right drawing, IFAS, Center for Aquatic Plants. University of Florida Gainesville (CAP/UFG); bottom left photo, MDNR.

p. 50 Two left drawings, Karen Teramura, USFWS/DOI; drawing second from right, Rebecca A. Haefner (see full credit for p. 19, above); right top and bottom drawings, IFAS CAP/UFG; bottom left and second from left, MDNR; bottom third from left, Peter W. Bergstrom, NOAA/DOC.

p. 51 Left drawing, John Norton; three right drawings, Karen Teramura, USFWS/DOI; bottom left photo, Peter W. Bergstrom, NOAA/DOC; bottom second and third from left, Linda M. Hurley, USFWS/DOI; bottom right, MDNR.

p. 52 Left and right drawings, Karen Teramura; middle drawing Ronald C. Phillips (see full credit for p. 47, at left); bottom left photo, Linda M. Hurley, USFWS/DOI; middle, Ronald C. Phillips, UHM; right, Peter W. Bergstrom, NOAA/DOC.

p. 53 Left, Michael D. Naylor, MDNR; middle, Linda C. Hurley, USFWS/DOI; right, Wayne Carmichael (Wright State University), Mark Schneegurt (Wichita State University), and Cyanosite (www.cyanosite. bio.purdue. edu).

p. 54 Michael D. Naylor, MDNR.

p. 55 Drawing used with permission (in N.L. Britton and A. Brown. 1913. Illustrated Flora of the Northern States and Canada, Vol. 2:612. Dover Publications, Mineola, NY). Left and right photos, Michael D. Naylor, MDNR; middle photo, Peter W. Bergstrom, NOAA/DOC.

p. 56 Large photo and inset, Michael D. Naylor, MDNR.

p. 57 Drawing, Karen Teramura, USFWS/DOI; photo, Linda M. Hurley, USFWS/DOI.

p. 58 Large photo and inset, Michael D. Naylor, MDNR.

p. 59 Large and small photos, Michael D. Naylor, MDNR

p. 60 Drawing, John Norton; left photo, Justin T. Reel, USGS; right photo, Linda M. Hurley, USFWS/DOI.

p. 61 Kimberly L. Schulz, ASLO Image Library, www.aslo.org/education/library.html.

p. 62 Top left, Wayne Carmichael (Wright State University), Mark Schneegurt (Wichita State University), and Cyanosite (www.cyanosite. bio.purdue.edu); right, Michael D. Naylor, MDNR; bottom left, David Krogmann (Purdue University), Mark Schneegurt (Wichita State University); and Cynaosite (www.cyanosite. bio.purdue.edu).

p. 63 Mike Haramis, USGS/BRD.

p. 64 Graph, adapted from the Chesapeake Bay Program and Moore et al. (2004); photograph, Peter W. Bergstrom, NOAA/DOC.

p. 65 Left and right, Skip Brown; middle, National Resources Conservation Service, www.nrcs. usda.gov/

p. 66 Top, and bottom, Linda M. Hurley, USFWS/DOI.

pp. 67-70 Paula Jasinski and Tara Shleser, NOAA CBO, Gloucester Point, Virginia.

ABOUT THE AUTHORS

Peter W. Bergstrom is a fishery biologist at the National Oceanic and Atmospheric Administration (NOAA) Chesapeake Bay Office in Annapolis, Maryland. His work involves conducting and funding SAV restoration, monitoring, and research, and he co-chairs the Living Resources Analysis Workgroup of the Chesapeake Bay Program. His research interests include habitat requirements for SAV, changes in SAV distribution by species, and effects of episodic events (such as algae blooms and rapid changes in benthic populations) on SAV abundance. He holds a Ph.D. in evolutionary biology from the University of Chicago.

Robert F. Murphy serves as director for the non-profit research center, Ecosystem Solutions. Bob oversees technical aspects of aquatic restoration projects and the development of innovative habitat restoration practices. He previously served as senior-level staff at the Alliance for the Chesapeake Bay (ACB), where he supervised all ACB oyster and SAV restoration projects throughout the Chesapeake and in Maryland and Virginia's coastal bays. Bob's research interests include seagrass ecology, fish-habitat interactions, and human-induced alterations to marine systems. He holds a Masters degree in fish ecology from the University of Maryland.

Michael D. Naylor is a natural resources biologist at the Maryland Department of Natural Resources in Annapolis, Maryland. Mike has worked on issues surrounding aquatic plants, fisheries, and exotic species in his eleven years with the agency. For the last five years he has acted as chairman for the Chesapeake Bay Program's Submerged Aquatic Vegetation Task Group. Having spent thousands of hours in the field identifying aquatic plants throughout tidal and non-tidal Maryland, Mike is often called upon by scientists and citizens to identify confusing plant specimens.

Ryan C. Davis is a managing scientist with Quantitative Environmental Analysis, LLC in Glens Falls, New York. Ryan's research has focused on developing restoration methods and field sampling techniques to quantify the success of restored submerged aquatic vegetation (SAV) habitat. His project experience includes developing and implementing large-scale SAV habitat restoration programs in the Chesapeake Bay, New England, and New York. He holds a Ph.D. in natural resources from the University of New Hampshire.

Justin T. Reel is an environmental project scientist for Rummel, Klepper & Kahl, LLP (RK&K) in Baltimore, Maryland. Justin's work in the Chesapeake Bay has focused on large-scale SAV restoration projects and water quality monitoring in the Potomac River. His SAV restoration efforts have also included sites in the Piscataqua River and Little Harbor in New Hampshire. Prior to joining RK&K, Justin served as an environmental scientist with the U.S. Geological Survey. In this capacity, his focus was on water quality, and SAV distribution and abundance in the Potomac River.